MOLECULAR BIOLOGY TECHNIQUES

AN INTENSIVE LABORATORY COURSE

MOLECULAR BIOLOGY TECHNIQUES

AN INTENSIVE LABORATORY COURSE

Walt Ream and Katharine G. Field

Department of Microbiology
Oregon State University
Corvallis, Oregon

Academic Press

San Diego London Boston New York Sydney Tokyo Toronto

This book is printed on acid-free paper. ∞

Academic Press
a division of Harcourt Brace & Company
525 B Street, Suite 1900, San Diego, California 92101-4495, USA
http://www.apnet.com

Academic Press
24-28 Oval Road, London NW1 7DX, UK
http://www.hbuk.co.uk/ap/

Library of Congress Catalog Card Number: 98-87608

International Standard Book Number: 0-12-583990-1

PRINTED IN THE UNITED STATES OF AMERICA
98 99 00 01 02 03 EB 9 8 7 6 5 4 3 2 1

Contents

Preface

We designed this intensive laboratory course to teach incoming graduate students the basic skills of molecular biology, so that students with varied research experience would understand the fundamental principles. To do this, we created an intensive 2-week course that includes techniques commonly used in molecular biology—methods that provide the foundation for most other procedures. In this course students learn how to prepare and analyze DNA, RNA, and proteins. The class must meet all day to cover the material in a 2-week period. This approach has several advantages. Students learn that the laboratory is open nights and weekends, and they discover that the day is not done until they complete their work. The students focus all their attention on learning molecular biology, and they learn to perform several tasks at once.

Soon, faculty, senior graduate students, technicians, and postdoctoral fellows joined our class, and beneficial interactions occurred among participants at different stages of their careers. Word of our course spread, and now each session includes faculty and staff from major research universities, teaching colleges, and biotechnology companies from around the world. Professionals unable to devote an entire term to a course often can spend 2 weeks away from home.

We based our choice of protocols for this book on their reliability and instructional value. For example, we use cesium chloride–ethidium bromide density-gradient centrifugation to purify large quantities of plasmid DNA, even though this method is seldom used in most laboratories. This method allows students to observe the differences in topological forms of DNA, it yields DNA of higher purity than other procedures, and students learn how to use an ultracentrifuge, which is something few have done and many will need to do. We use radioactive probes in our course for similar reasons. Radioactive probes offer greater sensitivity, less background signal, and easier washing and detection than the nonradioactive methods we have tried. In addition, students learn to handle radioactive isotopes safely, which is something most graduate students must learn. Although in our current research we use polymerase chain reaction to create specific mutations, we still teach oligonucleotide-directed mutagenesis because it gives students hands-on experience with many concepts not covered elsewhere. Thus, this course includes seldom-used procedures with heuristic value together with methods used routinely in typical molecular biology laboratories.

Graduates of our course have successfully applied these approaches to their own experimental systems, and they have acquired the skills needed to teach themselves new procedures from *Current Protocols, Molecular Cloning,* and other sources. For example, one of our former students, William Proebsting, a professor of horticulture at Oregon State University, used skills he acquired in our class to clone the pea dwarfing gene studied by Mendel (*Proc. Natl. Acad. Sci. USA,* **94:** 8907–8911, August 1997). Thus, we have had the satisfaction of seeing this course prove beneficial to our students, and we have found it rewarding to teach. We hope you have the same experience.

Walter Ream
Katharine G. Field

Course Synopsis

INTRODUCTION

Molecular biology, in particular recombinant DNA research, has transformed research in the biological and medical sciences. This technology currently influences all aspects of biological research, has far-reaching applications in clinical diagnosis, and has led to important developments in agriculture and biotechnology.

This course provides a hands-on introduction to molecular biological methods, including molecular cloning, polymerase chain reaction (PCR), Southern (DNA) blotting, Northern (RNA) blotting, DNA sequencing, oligonucleotide-directed mutagenesis, and protein expression, purification, and detection. You will work with a well-characterized gene (*virD2*) from the *Agrobacterium tumefaciens* tumor-inducing (Ti) plasmid. Virulence (*vir*) genes mediate transfer of a specific region of the Ti plasmid from *A. tumefaciens* into host plant cells; oncogenes contained in the transferred DNA integrate into the host nuclear genome where their expression causes tumorous growth. Because we have studied *virD2* previously, both starting materials and finished products for each experiment are available. All of the experiments you will do are "real," and most are published. This approach will demonstrate practical aspects of experimental design. Although the gene you will use is bacterial, the techniques apply to any research system.

Cloning, Restriction Analysis, Protein Expression, and Western Blots

You will begin with a plasmid (pWR160) that contains a portion of the *virD* operon inserted in a commonly used *E. coli* vector plasmid (pUC18). This plasmid contains *virD1* and *virD2*, which encode a 2-component, site-specific endonuclease that initiates transfer of DNA into plant cells.

To purify VirD2 protein, you will fuse the *virD2* gene to *gst,* a gene encoding glutathione-*S*-transferase. The Gst–VirD2 fusion protein retains the ability to bind glutathione and can be purified many-fold by affinity chromatography on glutathione–Sepharose. An expression vector plasmid (pGEX2) contains *gst* downstream from a strong promoter (the *tac* promoter); restriction endonuclease cleavage sites lie at the 3' end of the *gst* coding sequence.

Before you construct the gene fusion, you must purify the cloning vector plasmid (pGEX2) and the template for PCR DNA synthesis (pWR160). You will extract plasmid DNA from *E. coli* cells that harbor these plasmids and use CsCl–ethidium bromide density gradient centrifugation to purify the plasmid DNAs.

To fuse *virD2* to *gst,* you will cleave the vector plasmid (pGEX2) with the restriction endonucleases *Bam*HI and *Eco*RI. Polymerase chain reaction allows you to add appropriate restriction sites to the ends of the *virD2* coding sequence so that you can fuse it, in the same reading frame, to *gst*. The PCR primers contain restriction sites (at their 5' ends) adjacent to sequences complementary to the ends of *virD2*. You will anneal these primers to denatured template DNA (pWR160) and use PCR to amplify a DNA fragment consisting of *virD2* with a *Bgl*II restriction site at the 5' end and an *Eco*RI site at the 3' end of the gene.

Next treat the PCR product (containing *virD2*) with restriction enzymes *Bgl*II and *Eco*RI. After you inactivate the restriction enzymes, mix the DNAs and incubate them with T4 DNA ligase to join their cohesive ends. After ligation, transform *E. coli* cells with the ligated DNA. The vector plasmid confers ampicillin resistance on its host, allowing you to select transformed *E. coli* cells. To analyze the plasmids contained in ampicillin-resistant transformants, grow small liquid cultures of individual transformants and use an alkaline lysis method to prepare plasmid DNA for restriction endonuclease digestion and agarose gel electrophoresis. Once you identify transformants that contain the *gst-virD2* gene fusion, you can express and purify the Gst-VirD2 fusion protein.

You will examine the Gst–VirD2 fusion protein using SDS-polyacrylamide gel electrophoresis and western blot immunological detection. To induce expression of the *gst–virD2* gene, grow *E. coli* containing this gene fusion in broth containing IPTG, which induces the *tac* promoter. Next, prepare crude extracts from these cells and allow the fusion protein to bind glutathione–Sepharose beads. Washes will remove most other proteins, resulting in a substantial purification of bound Gst–VirD2 protein. Examine the crude extracts and purified proteins on silver-stained SDS-polyacrylamide gels. You will also transfer proteins from gels to nylon filters and use antisera raised against Gst to detect the Gst–VirD2 fusion protein; this is the Western blot procedure.

Oligonucleotide-Directed Mutagenesis

You will create a specific mutation in *virD2* using an oligonucleotide primer that contains the mutation. To produce the single-stranded template, we will insert *virD2* sequences into a phagemid vector (pUC119). A phagemid contains origins of replication from a plasmid (ColE1) and a single-stranded DNA phage (M13); therefore, phagemid DNA can be isolated in either single- or double-stranded form. You will isolate uracil-containing single-stranded phagemid DNA from an *E. coli* strain deficient in dUTPase and uracil repair (*dut− ung−*). This DNA will serve as template for DNA synthesis (in vitro) primed by an oligonucleotide complementary (except for the desired mutation) to *virD2*. On completion, the uracil-containing template will be paired with newly synthesized thymine-containing DNA, except at the short heteroduplex region created by mismatches in the mutagenic primer oligonucleotide. On transformation into a uracil repair-proficient (*ung*+) *E. coli* strain, the uracil-containing wild-type template will be degraded, and the thymine-containing mutant strand will give rise to transformants containing the mutation in *virD2*. You will confirm putative mutants by restriction analysis and DNA sequencing.

DNA Sequence Analysis

In vitro DNA synthesis can introduce unwanted mutations into DNA. To determine whether your mutant *virD2* contains mutations in addition to the oligonucleotide-directed mutation, you will determine its nucleotide sequence.

Southern (DNA) and Northern (RNA) Blotting

For Southern (DNA) blotting, you will use radiolabeled plasmid DNA containing *virD2* (pWR160) as a hybridization probe to examine genomic DNAs from *A. tumefaciens* for *virD2*. You will examine restriction fragment length polymorphisms (RFLPs) between *virD2* genes from different strains of *A. tumefaciens* and *A. rhizogenes.*

To examine the *virD2* region for RFLPs, you will isolate genomic DNAs from different strains of *A. tumefaciens,* digest these DNAs with restriction endonucleases, and separate the resulting fragments according to size by agarose gel electrophoresis. You will then denature the DNA and transfer it to a nylon filter. To the filter-bound DNA you will hybridize probe (pWR160) DNA labeled with ^{32}P by nick translation. Finally, you will wash the filter and detect hybridization by autoradiography.

Because bacterial messages are extremely difficult to detect by the Northern (RNA) blot procedure, we will depart from the sequence of experiments involving *virD2*. Instead, northern blot analysis will detect mRNA encoded by the *rbsS* (RuBisCO) gene among total RNA isolated from tobacco leaves. After extracting RNA from leaves, you will separate RNA molecules according to size by formaldehyde–agarose gel electrophoresis, transfer the RNA to a nylon filter, hybridize labeled probe DNA to the filter-bound RNA, wash the filter, and detect hybridization by autoradiography. The hybridization probe will be a pUC-based plasmid (pRbsS) containing part of the *rbsS* gene.

Protein Interaction Analysis in Yeast

You will use the yeast two-hybrid protein interaction assay to determine whether two *Agrobacterium* virulence proteins—VirE1 and VirE2—interact. The *virE2* gene is fused to a portion of *E. coli lexA* that encodes the sequence-specific DNA-binding domain of the LexA repressor. *virE1* is fused to a segment of the *Saccharomyces cerevisiae GAL4* gene that encodes the transcription activation domain of this yeast transcription factor. If VirE1 and VirE2 interact in vivo, the interaction will tether the Gal4 transcription activation domain to a LexA binding (operator) site located upstream from a reporter gene, which consists of the yeast *GAL1* promoter fused to the *E. coli lacZ* (β-galactosidase) gene. If *lacZ* is induced, the resulting β-galactosidase will turn the yeast blue in the presence of the chromogenic substrate X-gal. Controls needed to interpret the two-hybrid analyses are included.

SAFETY PRECAUTIONS

The following safety tips address common hazards encountered in molecular biology laboratories.

1. **Wear safety glasses when working with phenol and other caustic solvents,** including concentrated acids and bases.
2. **Never put bottles containing phenol or other caustic solvents on shelves over a lab bench.**
3. **Disconnect leads from power supplies after use;** capacitors can retain lethal electric charges, even when the unit is unplugged. When power supplies are in use, secure leads on the bench, not hanging down in front. To shut off power supplies, turn the rheostat to "0" first, then switch the power off. This will prevent the next user from blowing the unit's fuse, which can occur if the power supply is set to deliver full power the instant it is turned on.
4. **Put sharps in designated receptacles.** Razor blades, Pasteur pipets, needles, and other sharp objects will injure custodians if placed in the regular trash.
5. **Ethidium bromide (EtBr)** is a carcinogen. Wear gloves when handling it, and dispose of it in designated receptacles.
6. **Wear gloves and a lab coat when handling radioisotopes.** Use ^{32}P behind shields and limit body exposure to it. Scan hands, equipment and work areas after use. Dispose of radioactive material in radioactive waste receptacles.
7. **Autoclave materials containing bacterial cultures.** Dispose of culture plates in designated autoclave bags.

8. **Wear a lab coat.** It will protect you from chemical or isotopic contamination. Short pants and sandals are not safe attire in a laboratory.
9. **Use mechanical pipettors; avoid contact with reagents.**
10. **Do not eat or drink in the laboratory.**
11. **Protect eyes and skin from ultraviolet light.**

DAILY SCHEDULE

Day 1

8:00	Lab lecture
9:30	**I.A, steps 2–19.** Cesium chloride–ethidium bromide density gradient centrifugation: purification of plasmids; spin takes 4 hours
12:00	*Lunch break*
1:00	Lecture: Plasmids and Cosmids; Gene Cloning
2:30	Supplementary Lecture I: Gene Cloning II: Libraries
3:30	**I.A, steps 20–24.** Collect plasmid DNAs from the cesium chloride gradients, extract, precipitate, and quantitate

Day 2

8:00	Lab lecture
8:45	**I.C, steps 1–4.** Restriction digest of pGEX2 **III.A, steps 1–4.** Restriction digest of pCS64 and pUC119 **I.B, steps 1–8.** PCR to synthesize *virD2* flanked with restriction sites **I.D and III.B, steps 1–14.** Agarose gel and Geneclean restriction fragments
12:00	*Lunch break*
1:00	Lecture: Gene Amplification, RAPDs, Sequencing
2:00	**I.B, steps 9–11.** Agarose gel of PCR products **I.C, steps 5–7.** Restriction digest of PCR products **V.A, steps 1–17.** Prepare *A. tumefaciens* genomic DNAs **I.E, steps 1–5, III.C, steps 1–10.** Ligate DNA fragments to vectors

Day 3

8:00	Lab lecture
8:45	**V.B, steps 1–3.** Restriction digest of genomic DNAs
9:15	Supplementary Lecture II: Microbiological Techniques
9:45	**I.F, steps 1–12, III.D.** Transform *E. coli* DH5α with ligated plasmids
12:00	*Lunch break*
1:00	Lecture: Probes, Southern Blots, RFLPs
2:00	**V.C, steps 1–5.** Agarose gel of restriction fragments from genomic DNAs **V.D, steps 1–11.** Denature DNA and transfer to filter by blotting

Day 4

8:00	Lab lecture
8:45	**V.D, step 12.** Wash and UV cross-link DNA blot **V.F, step 1.** Prehybridization of Southern blot **V.E, steps 1–11.** Prepare probe **VI.A, steps 1–8.** Prepare total RNA from tobacco
12:00	*Lunch break*
1:00	Lecture: Cloning by Function, Two-Hybrid Screening
2:00	**V.F, step 2.** Hybridize Southern blot to probe **I.F, steps 13–14, III.D.** Pick colonies from transformations, inoculate broth

Day 5

8:00 Lab lecture

8:45 **I.G, steps 1–11, III.E.** Prepare plasmid DNA
V.F, steps 3–5. Wash Southern blot

10:15 **III.G, steps 3–6.** Template preparation: inoculate, add ampicillin, add helper phage
I.H, steps 1–13, III.F. Restriction analysis of recombinant plasmids
VI.A, steps 9–12. Precipitate RNA, dissolve, and reprecipitate

12:00 *Lunch break*

1:00 Lecture: Protein Isolation, Fusion Proteins, Expression Vectors

2:00 **I.H, steps 4–10, III.F.** Restriction analysis of recombinant plasmids: gels
III.G, step 7. Template preparation: add ampicillin and kanamycin, incubate
V.F, steps 6–7. Southern blot: autoradiography
VI.A, steps 13–15. Precipitate RNA, dissolve and quantify

Day 6

8:00 Lab lecture

8:45 **VI.B, steps 1–2.** Pour agarose formaldehyde gel for Northern blot

9:45 **VI.B, steps 3–6.** Formaldehyde-agarose gel electrophoresis of RNA

11:30 **VI.B, steps 7–8, VI.C, steps 1–11.** RNA transfer by blotting

12:00 *Lunch break*

1:00 Lecture: Northern Blots and Other RNA Techniques

2:00 **VI.D, steps 1–11.** Prepare probe
VI.C, step 12. Rinse and cross-link RNA blot
III.G, steps 8–9. Template preparation: centrifuge, begin PEG precipitation

Day 7

8:00	Lab lecture
8:45	**III.G, steps 10–15.** Finish single-stranded template preparation **III.H, steps 1–5.** Phosphorylate oligonucleotide primer
12:00	*Lunch break*
1:00	Lecture: Oligo-Directed Mutagenesis
2:00	**III.I, steps 1–4.** Anneal primer to template **III.J, steps 1–3.** DNA synthesis by primer extension **III.K, step 1.** Transform synthesis reactions into *E. coli* DH5α **VI.E, step 1.** Begin Northern blot prehybridization; incubate overnight

Day 8

8:00	Lab lecture
8:30	**VII.A.** Transform yeast **III.K, step 2.** Pick colonies, inoculate broth
10:45	**II.B, steps 1–2.** Pour resolving gel
11:30	**II.A, step 2.** Induce with IPTG, incubate
12:00	*Lunch break*
1:00	Lecture: In situ Hybridization
2:00	**II.B, steps 3–4.** Pour stacking gel
2:30	**II.A, steps 3–11.** Purify fusion protein on glutathione sepharose
3:00	**II.B, steps 5–8.** SDS-polyacrylamide gel electrophoresis **II.C, step 1–2.** Store gel overnight in gel fix **VI.E, step 2.** Begin Northern hybridization; incubate overnight

Day 9

8:00	Lab lecture
8:30	**III.L.** Minipreps of plasmid DNA
9:30	**III.M, step 1.** Restriction digest of plasmid DNA with *Hind*III and *Bam*HI
10:30	**II.C, steps 1–13.** Silver staining of protein gel **II.B, steps 1–3.** Pour resolving part of SDS-polyacrylamide gel; store in refrigerator
12:00	*Lunch break*
1:00	Lecture: Computer Analysis of Gene Sequence Data
2:00	Group 1: Computer analysis of nucleic acid sequences Group 2: **III.M, steps 2–8.** Gel of restriction digest to confirm mutants **VI.E.** Northern blot washes
3:30	Group 1 switch with Group 2

Day 10

8:00	Lab lecture
8:45	**IV.A, steps 1–3.** Pour gels for sequencing **IV.B, steps 1–13.** Sequencing reactions **IV.A, steps 4–6.** Sequencing gel electrophoresis
12:00	*Lunch break*
1:00	Lecture: Methods for Analysis of Gene Expression
2:00	**II.B, steps 4–8.** Pour stacking gel; perform electrophoresis **IV.A, steps 7–8.** Dry sequencing gels, put on film **II.D, steps 1–4.** Electrophoretic transfer of proteins **II.D, steps 5–9.** Antibody incubation of Western blot
3:30	Lecture: DNA-Protein Interactions

Day 11

8:00	Lab lecture
8:45	**II.D, steps 10–15.** Secondary antibody incubation, color development **IV.** Develop films from Northern blot, examine, and discuss **VII.B.** β-galactosidase assays
12:00	*Lunch break*
1:00	Lecture: Antibodies, Immunological Techniques
2:00	Develop all films; examine and discuss

Lab Cleanup

ACKNOWLEDGMENTS

We thank Dr. Gary Merrill, Dr. Meredith Howell, and the many outstanding teaching assistants who contributed to the development of this course. The following companies donated materials to support this course: Amresco, Fisher, GIBCO/BRL Life Technologies, ICN, New England Biolabs, Perkin Elmer, Rainin, RPI, and Upstate Biotechnologies.

EXERCISE 1

DNA Preparation, Polymerase Chain Reaction, and Molecular Cloning

Background

The goal of this experiment is to fuse *virD2* (a virulence gene from *Agrobacterium tumefaciens*) to *gst* (glutathione-*S*-transferase gene) to express and purify the fusion protein. To accomplish this, you need to create restriction sites on either side of *virD2*. You will use polymerase chain reaction (PCR) to make copies of *virD2* flanked by the required restriction sites. These restriction sites will allow you to insert *virD2* into the expression vector pGEX2. To start this and subsequent experiments, you need several different plasmids, which you will purify.

Steps of the experiment are

A. Prepare plasmid by cesium chloride density gradient centrifugation

B. Use PCR to synthesize *virD2* flanked with restriction sites

C. Perform restriction digests of plasmid pGEX2 and PCR products

D. Purify DNA fragments from agarose

E. Ligate PCR product to pGEX2

F. Transform *E. coli* with ligated plasmid

G. Make small-scale preparation of plasmid DNA from broth cultures

H. Perform restriction digests of DNAs; examine to confirm insert

A. CESIUM CHLORIDE–ETHIDIUM BROMIDE DENSITY GRADIENT CENTRIFUGATION

Introduction

The purpose of CsCl–ethidium bromide density gradient centrifugation is to separate supercoiled (covalently closed circular) plasmid DNA from other components present in lysates of bacterial cells: linear DNA (chromosomal fragments), nicked plasmid DNA (which lacks supercoils), RNA, proteins, and carbohydrates. Covalently closed circular plasmid DNA isolated from bacterial cells has one strand underwound (fewer helical turns per unit length) relative to the other. To relieve the strain caused by underwinding, the circular DNA twists into figure eight structures called negative supercoils. The axis of the DNA double helix crosses over itself multiple times, much like an overwound telephone cord.

After cell lysis, low-speed centrifugation removes debris and much of the chromosomal DNA from the lysate. High-speed centrifugation in the presence of CsCl and ethidium bromide separates components on the basis of their buoyant densities. During centrifugation, the CsCl solution forms a density gradient. Ethidium bromide is a planar molecule that intercalates between the bases in DNA. As ethidium intercalates into a negatively supercoiled plasmid, the superhelicity is relaxed; at this point, intercalation of additional ethidium molecules must introduce positive supercoils into the plasmid DNA, but this process requires energy. Therefore, closed circular plasmids intercalate less ethidium than linear DNA or nicked (relaxed) circular DNA. Because ethidium is less dense than DNA, the density of linear DNA–ethidium is less than that of closed circular DNA–ethidium; this difference in density provides the basis for separating these topological forms. (RNA, which also binds ethidium, is more dense than DNA and forms a pellet, whereas protein is less dense than DNA.)

This procedure differs from most CsCl gradient protocols because it requires 10-fold less ethidium bromide. Use of such a small quantity of ethidium bromide is only possible on removal of ethidium-binding RNA by ribonuclease digestion. The DNA bands will appear pink against a colorless background (in subdued room light). Gradients that contain typical concentrations of ethidium will appear orange, and bands become visible only under UV light (which damages DNA). DNA prepared from lysates treated with RNase will contain significant ribonuclease activity. This is a problem only if the DNA is used for in vitro transcription, but phenol extraction will destroy the RNase activity.

We will prepare the following plasmid DNAs:

pWR160:	contains *virD2*; template for PCR synthesis (I); hybridization probe for Southern blot (V)
pGEX2:	expression vector plasmid (Pharmacia) (I and II)
pUC119:	phagemid vector for mutagenesis (III)
pCS64:	source of *virD2* restriction fragment; target for mutagenesis (III)
pMCB525:	oligonucleotide-directed mutagenesis template (III)
pRbcS:	contains *rbcS;* hybridization probe for Northern blots (VI)
pAD	yeast plasmid with transcription activator (TA) domain (VII)
pAD–E1	yeast plasmid with TA–*virE1* fusion (VII)
pBD	yeast plasmid with *lexA* DNA-binding domain (VII)
pBD–E2	yeast plasmid with *lexA–virE2* fusion (VII)

Before beginning any experiment involving bacterial cultures, streak the strain on the appropriate agar plates to obtain single colonies, assuring a pure culture. Then inoculate liquid medium (usually a rich broth, containing antibiotics when appropriate) with a single colony, and incubate the culture with aeration at the proper temperature (usually 37°C for *E. coli*) overnight.

Safety Precautions

Ethidium bromide is mutagenic; avoid contact with it. Wear gloves.

Discard butanol in the organic waste bottle in the hood.

Technical Tips

RNases are extremely stable enzymes and are resistant to heating. Their presence will destroy RNA preparations. To avoid RNase contamination during future RNA work, use RNase only in one area under the hood and pipet RNase with the pipettor designated for RNase work.

Protocol

Beforehand (TAs)

1. Inoculate 200 mL of L broth (containing ampicillin at 50 μg/mL) with 1 mL of fresh overnight L broth culture of an *E. coli* strain harboring the desired plasmid. Incubate with aeration at 37°C overnight.

Day 1 (students start here)

2. Place 200 mL cells from an overnight culture in a 250-mL centrifuge bottle. Balance bottles and centrifuge cells 10 minutes, 5000 rpm, 4°C in Sorvall GSA.
3. While cells are in centrifuge, prepare fresh SDS–NaOH lysis buffer (200 mM NaOH; 1% SDS; mix

0.67 mL of 3 *M* NaOH + 0.4 mL of 25% SDS + 9 mL H_2O). Do not put lysis buffer on ice, or SDS will precipitate.

4. **Thoroughly** resuspend cells from 100 to 200 mL of culture in 2 mL of TE (25/10). Transfer to a sterile 50-mL "Oakridge" Nalgene tube.
5. Add 4 mL of SDS–NaOH lysis solution; mix **gently but thoroughly by inverting tube several times** (lyses cells). Incubate on ice 5 minutes and keep on ice from here on.
6. Add 3 mL of potassium acetate solution; mix **gently but thoroughly** (neutralizes).
7. Centrifuge 13,000 rpm, 15 minutes, 4°C, in Sorvall SS34 (pellet contains cellular debris and chromosomal DNA).
8. Decant supernatant into weighed 50-mL "Oakridge" Nalgene tube.
9. Add 16 mL 95% ethanol to supernatant.
10. Centrifuge 10,000 rpm, 10 minutes, 4°C, in Sorvall SS34 (precipitates plasmid DNA).
11. During centrifugation, weigh 4.5 g of CsCl into a 12-mL Falcon tube.
12. Discard supernatant. Dissolve pellet in 3 mL TES buffer.
13. Add 5 μL of 5 mg/mL heat-treated RNase.
14. Add TES to bring resuspended pellet to 4 g.
15. Add 4.5 g of CsCl.
16. Use a Pasteur pipet to load solution in Beckman polyallomer 13- × 51-mm heat-seal tube (No. 342412).

17. Layer 30 μL of 10 mg/mL ethidium bromide on top of CsCl solution; do not mix yet; fill to base of neck with H_2O.

18. Weigh to find tubes that balance. Tube should weigh about 9.6 g. Balance tubes within 100 mg. Heat-seal tube; check seal; weigh again.

19. Centrifuge 65,000 rpm, 4 hours, 15°C, in Beckman vTi80 rotor.

20. Collect plasmid DNA band with 18-gauge syringe needle (this procedure will be demonstrated; also see diagram). Draw DNA into syringe and discard needle **before** you expel DNA into a 1.5-mL microfuge tube; this reduces shearing. To avoid spreading RNase around the lab, prepare a receptacle for the excess solution **before** you puncture the tube.

21. Extract ethidium 4 to 5 times with H_2O-saturated *n*-butanol. To extract, add 1 volume butanol, vortex, spin 2 seconds in microfuge. Butanol layer will be on top. Remove the butanol and repeat the extraction 3 or 4 more times.

22. Dilute with 2 volumes TE (10/0.1); if your sample volume is now greater than 500 μL divide it between 2 tubes. Add 2 volumes of ethanol (based on the volume of sample + TE together). Hold on ice for at least 10 minutes.

23. Centrifuge in microfuge at top speed for 5 minutes. Thoroughly remove supernatant and air dry pellet to evaporate remaining ethanol. If you divided your sample among different tubes, take up each pellet in a smaller volume and combine.

24. Measure optical density (OD) at 260 nm; concentration (μg/mL) = $OD_{260\ nm}$ × dilution factor × 50. To measure OD, prepare 100 μL of a dilution of your sample, using 90 μL buffer and 10 μL sample for a

dilution factor of 10. Zero the spectrophotometer with the buffer used to prepare your dilution.

Solutions for Cesium Chloride Density Gradient Centrifugation

TE (25/10): 25 mM Tris, pH 7.5; 10 mM EDTA (autoclave)
SDS-NaOH lysis solution: 200 mM NaOH; 1% SDS
Mix 0.67 mL of 3 *M* NaOH + 0.4 mL of 25% SDS + 9 mL H_2O
Potassium acetate solution: 3 *M* K+/5 *M* CH_3CO_2- (autoclave)
Mix 60 mL of 5 *M* potassium acetate (29.45 g/60 mL) + 11.5 mL of acetic acid + 28.5 mL of H_2O
TE (10/0.1): 10 mM Tris, pH 7.5; 0.1 mM EDTA (autoclave)
TES: 50 mM Tris, pH 8.0; 5 mM EDTA; 50 mM NaCl (autoclave)
Ethidium bromide stock: 10 mg/ml in H_2O (store in dark) (avoid contact with this mutagen)
RNase solution: mix 5 mg/mL RNase A + 5 mg/mL RNase T1 in DNA buffer; heat at 15°C for 15 minutes (to inactivate DNases)
H_2O-saturated *n*-butanol: shake equal volumes of water and *n*-butanol
95% Ethanol
4 mL TES and 4.5 g CsCl (solution to top off tubes)
(Bacterial strains and plasmids are available from Walt Ream, Dept. of Microbiology, Oregon State University, Corvallis, OR 97330. Phone: 541-737-1791 E-mail: reamw@bee.orst.edu.)

IDENTIFICATION OF NUCLEIC ACID BANDS AFTER VERTICAL ROTOR CENTRIFUGATION

If pretreated with RNase | If RNase treatment omitted

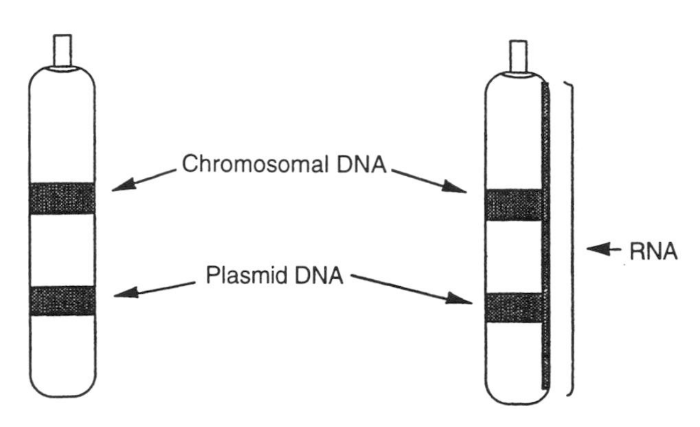

Procedure

1. Insert needle (without syringe) into top of tube; leave in place.
2. Insert needle (with syringe attached) through wall of tube 0.5 cm below the plasmid DNA band. **Do not** twist the needle; push it straight in with steady pressure. (Puncture the tube the way you hope the nurse does when drawing blood from your arm.) Keep the bevel up; place the bevel at the bottom of the band and draw the plasmid band into the syringe.

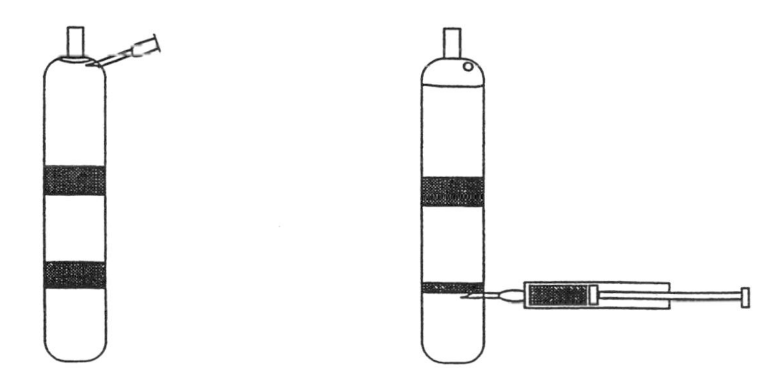

Notes

B. PCR TO SYNTHESIZE *VIRD2* FLANKED WITH RESTRICTION SITES

Introduction

Polymerase chain reaction produces many copies of a particular template DNA sequence in vitro. This method takes advantage of thermally stable DNA polymerase, which is not denatured by temperatures high enough to melt apart duplex DNA, to allow repeated cycles of thermal template denaturation, primer annealing, and DNA synthesis. In theory, 30 cycles of amplification can produce 268,435,456 copies from a single template molecule. Because the yield rises geometrically with each cycle, it is extremely sensitive to factors that affect efficiency of the reaction (e.g., pH, hybridization temperature, magnesium ion concentration). PCR is also extremely sensitive to contamination by unwanted template DNA in reagents.

The reaction mix contains template DNA, polymerase, deoxynucleoside triphosphates, Mg^{2+}, buffer, and 2 primer oligonucleotides. One primer complements a site *upstream* from the sequence being amplified; the other primer complements a region on the *opposite* strand, *downstream* from the sequence. Primers are typically 18 to 30 bases long. The reaction temperature is raised to denature the template DNA (94°C), reduced to allow the primers to anneal to their target sequences (usually 45 to 55°C), and then raised to the optimum temperature for DNA polymerase activity (72°C). This cycle is repeated, usually 25 to 40 times. During the first cycle, the primers anneal to the target sequence and are extended by DNA polymerase beyond the other primer-binding site. In subsequent cycles newly extended strands, which start at either side of the gene, are used as templates, generating short strands that extend from primer to primer.

An additional sequence, such as a restriction endonuclease recognition site, can be added as a "tail" at the 5' end of a primer, provided the sequence complementary to the template is maintained. In our experiment, the 3' end of the left-hand primer (5'-CCC CTG ATC AAG ATC TAG ATC TAT GCC CGA TCG CGC TCA A-3') complements the first 18 bases of the *virD2* antisense strand. The 5' end of this primer contains two *Bgl*II restriction sites (AGATCT) and one *Bcl*I site (TGATCA). Similarly, the 3' end of the right-hand primer complements the last 18 bases of the *virD2* sense strand, and an *Eco*RI site (GAATTC) lies at its 5' end. The purpose is to produce the *virD2* coding sequence with restriction sites at both ends, allowing you to cut the PCR product and fuse it to the expression vector.

You will use a commercial PCR kit, which contains the buffer, deoxynucleoside triphosphates, and *Taq* DNA polymerase (from *Thermus aquaticus*), as well as primers and template for a positive-control reaction.

Safety Precautions

Protect eyes and skin from ultraviolet light; avoid contact with ethidium bromide.

Technical Tips

A single molecule of contaminating DNA can be amplified many times; thus, PCR requires unusual care to avoid contamination. Discard pipet tips after each transfer; a common source of contaminated solutions is "double-dipping". Use pipet tips with barriers to avoid contaminating the pipettors, reagent stocks, and reactions. Always include negative control reactions with no added template DNA to check for contaminating DNA; perform an equal number of negative controls and template-containing reactions.

Practice opening and closing an empty thin-walled PCR tube to learn how to handle them without crushing.

Protocol

1. Turn on a heating block; set to 95°C.
2. Fill out a DNA Amplification Checklist (*see page* 32) for your reactions. Include a positive control with the primers and template supplied with the PCR kit, a negative control with no template DNA but all other reactants, a positive control using previously prepared pWR160 template provided by the TAs, and the experimental amplification using your prepared pWR160 template.

   These are the amounts for each amplification reaction:

   Reaction Cocktail:

10× buffer		10 μL
dNTPs, 2 mM stock		10 μL
Primer 1, 10 μM stock		2 μL
Primer 2, 10 μM stock		2 μL
Template DNA	10 ng =	μL
Taq polymerase	2.5 U –	0.5 μL
ddH_2O to bring total volume to 100 μL		μL

3. In 0.2-mL PCR tubes, add all the reactants except the *Taq* polymerase, starting with the water and then following the order listed above. Mix well.
4. Heat the reaction tubes to 95°C for 2 minutes.
5. Pulse the tubes in the microfuge.

6. Add the polymerase; mix without introducing bubbles.
7. Overlay the reaction with 2 drops of sterile mineral oil (omit oil if thermal cycler has a heated cover).
8. Place the tubes in the thermal cycler and run the following program:
   94°C for 1 minute; 55°C for 1 minute; 72°C for 3 minutes; 30 cycles; hold at 4°C (can be left overnight)
9. To estimate the yield of PCR product ("amplicon"), examine 5 μL of each reaction by agarose gel electrophoresis. Cast a 1% agarose gel (0.3 g agarose in 30 mL 1× TAE; heat in microwave until dissolved; replenish evaporated volume with distilled H_2O; add 1 μL of ethidium bromide; pour and insert comb).
10. Draw off 5 μL of each reaction from under the oil with a Pipetman and place in fresh tubes.
11. Add 1 μL of 6× load buffer; stir.
12. Prepare a size standard using 2 μL of DNA Mass Ladder, 2 μL ddH_2O, and 1 μL of 6× load buffer.
13. When gel has solidified, place in gel apparatus, add electrophoresis buffer, remove comb, and load. When loading an agarose gel, insert only the tip of the pipet tip into the well; you may pierce the bottom of the well if you insert the tip too far.
14. To turn on the power supply set rheostats to "0", turn on the power, and allow to warm up for a few minutes. Connect the leads to the gel apparatus and apply 100 V until bromphenol blue dye nears the bottom of the gel (approximately 30 minutes). DNA migrates toward the positive (red) electrode (anode). Set rheostats to "0" before you turn off the power supply. Examine under UV light (use a hand-held UV lamp and safety glasses) and continue electrophoresis as needed.
15. Photograph the gel using the UV transilluminator/ Polaroid camera apparatus. Please note that each

piece of film costs $1.25; take **1 picture only.** Make additional copies on the photocopier.

16. Estimate the yield of PCR reactions by comparing band intensity with the DNA Mass Ladder bands.
17. To remove mineral oil from the PCR product, withdraw DNA from under the oil with a pipet, place in fresh tubes, and extract with chloroform. Add 1 volume chloroform, vortex, and spin 2 minutes in microfuge. Remove the aqueous (top) layer to a fresh tube.

Solutions for PCR

10× PCR buffer (supplied by Perkin Elmer)
 100 mM Tris-HCl, pH 8.3, at 25°C
 500 mM KCl
 15 mM $MgCl_2$
 0.1% (w/v) gelatin

dNTPs: Deoxynucleoside triphosphates dissolved in ddH_2O and neutralized with NaOH. Stock solution contains 2 mM of each dNTP (in ddH_2O).

TAE: 0.04 *M* Tris-acetate, 0.002 *M* EDTA
 1 L 50×: 242 g Tris base
 57.1 mL glacial acetic acid
 100 mL 0.5 *M* EDTA, pH 8.0

Ethidium Bromide stock = 10 mg/mL; store in dark at 4°C.
 Caution: mutagen; wear gloves and a mask while weighing; wear gloves while using.

6× Load buffer: 0.05% bromphenol blue;
 40% (w/v) glycerol in H_2O (GIBCO/BRL)

DNA Mass Ladder (GIBCO/BRL)

Chloroform

Sterile mineral oil

Primer 1: 5'-CCC CTG ATC AAG ATC TAG ATC TAT GCC CGA TCG CGC TCA A-3'

Primer 2: 5'-CCC GAA TTC TAG GTC CCC CCG CGC CC-3'

DNA AMPLIFICATION CHECKLIST

DATE ______________ NAME ______________

No.	Template	H_2O	10× buffer	dNTPs	#1 primer	#2 primer	temp. DNA			*Taq* pol

Denaturation temp ____

time ____

Annealing temp ____

time ____

Elongation temp ____

time ____

Number of cycles ____

NOTES:

GEL PHOTO HERE

Notes

Notes

C. RESTRICTION DIGESTS OF PLASMID pGEX2 AND PCR PRODUCTS

Introduction

Type II restriction endonucleases occur mainly in prokaryotes; these enzymes recognize and cleave specific sequences in double-stranded DNA. Type II restriction enzymes require Mg^{2+}, salt (often NaCl, sometimes KCl), and buffer for activity, but they do not require ATP.

Restriction endonucleases make double-stranded cuts at specific sequences in the phosphodiester backbone of DNA. Their recognition sequences usually contain 4 to 6 base pairs and are palindromes (they read the same, 5' to 3', on either strand). Most restriction endonucleases bind and cut DNA within the recognition sequence. For example, *Bam*HI recognizes

5'-G/GATC C-3'
3'-C CTAG/G-5'

and makes staggered cuts at the slashes. Note that the recognition sequence reads 5'-GGATCC-3' on each strand (a palindrome) and the cuts are staggered, leaving 4 bases unpaired at the 5' end:

5'-G -3'
3'-CCTAG-5'

This is called a 5' extension or overhang. Because of these unpaired (cohesive) ends, the termini of any *Bam*HI restriction fragments can pair with each other. These cohesive ends allow us to join any *Bam*HI fragments together in vitro (using an enzyme called DNA ligase). **This is the basis for molecular cloning.**

To perform a restriction digest, start with 0.5 to 1.0 μg of the DNA you will cut per 20 μL of reaction volume.

Each restriction enzyme comes with its reaction buffer. Manufacturer's instructions will indicate the buffer to use for simultaneous digestion with 2 enzymes.

You will cut the expression vector plasmid pGEX2 with restriction enzymes *Bam*HI (G/GATCC) and *Eco*RI (G/AATTC). These enzymes leave incompatible ends, so that completely cut plasmid will not recircularize. To check the efficiency of each enzyme, cut pGEX2 with each enzyme separately. Examine uncut, single-cut, and double-cut plasmid by agarose gel electrophoresis.

In a separate digestion, cut the PCR product with *Bgl*II (A/GATCT) and *Eco*RI, which will leave it with ends compatible with those of the cut plasmid.

Technical Tips

The volume of restriction endonucleases added must not exceed 10% of the total reaction volume because glycerol in the enzyme preparations inhibits their activity at high concentrations.

Keep enzyme stocks chilled (store at −20°C; not in frost-free freezer). Keep restriction enzymes on ice for the short period of time that you have them out of the freezer. Add them last to a restriction digest mixture, and then begin the incubation immediately.

Protocol

Digestion of pGEX2 plasmid

1. Mix: 2 μg of vector (pGEX2) DNA (from your CsCl preparation)
   2 μL of 10× restriction buffer (NEBuffer *Eco*RI)
   distilled water to give 18 μl total volume

2. Add 5 to 10 units (usually 0.5 to 1.0 μL each) of the restriction endonucleases (*Eco*RI and *Bam*HI).

3. Incubate at 37°C for at least 2 hours.

4. As controls, set up 3 more reactions, but omit both enzymes in one and add only 1 of the 2 restriction enzymes to the others (*Eco*RI to one, *Bam*HI to the other). Incubate as above.

Hint: To save time when setting up multiple restriction digests, prepare a "master mix". For example, for these 4 reactions, combine:

8 μg of vector (pGEX2) DNA (from your CsCl preparation)
8 μL of 10× restriction buffer (NEBuffer *Eco*RI)
distilled water to give 76 μL total volume

Divide this into 4 tubes and add the restriction enzymes to each (5 to 10 units of *Eco*RI and *Bam*HI to the 1st, *Eco*RI to the 2nd, *Bam*HI to the 3rd, water to the 4th).

Digestion of PCR product

5. After you estimate the yield of PCR product,
Mix: 2 μg of insert DNA (your PCR product in this case)
2 μL of 10× restriction buffer (NEBuffer *Eco*RI)
distilled water to give 18 μL total volume

6. Add 10 units each (usually 1 μL) of the restriction endonucleases (*Eco*RI and *Bgl*II for our experiment).

7. Incubate at 37°C for 1 to 2 hours.

Solutions for Restriction Digests

NEbuffer *Eco*RI: 50 mM NaCl, 100 mM Tris–HCl, 10 mM $MgCl_2$, 0.025% Triton X-100 (pH 7.5 at 25°C)

Notes

D. PURIFICATION OF DNA FRAGMENTS FROM AGAROSE

Introduction

Many vector plasmids, for example pUC119, are designed so that insertion of a foreign DNA fragment disrupts a gene with an easily detectable phenotype, for example *lacZ* (β-galactosidase). This enzyme converts colorless X-gal (5-bromo-4-chloro-3-indolyl-β-D-galactoside) to a dark blue indigo derivative. When included in selective agar plates, X-gal allows us to distinguish Lac+ transformants (blue colonies) that received a recircularized vector plasmid (with no foreign DNA inserted) from Lac− transformants (white colonies) that contain an insert in the vector plasmid.

Unfortunately, pGEX2 does not provide a marker that allows us to distinguish cells transformed with uncut or recircularized vector from those harboring plasmids with foreign DNA inserted. Because we cleaved pGEX2 with enzymes (*Eco*RI and *Bam*HI) that produce incompatible ends, recircularization of the vector will occur infrequently. However, if the digested vector sample contains uncut pGEX2 (even small quantities not detectable by agarose gel analysis), these uncut molecules will transform *E. coli* very efficiently and reduce the likelihood that the transformants you analyze contain plasmids with PCR product inserted.

To avoid these problems, you will use agarose gel electrophoresis to separate linear vector DNA from uncut plasmid. Excise the portion of the gel containing cut vector, dissolve the agarose in sodium iodide, and bind the DNA to glass beads in high-salt buffer. Wash the beads to remove contaminating material, and elute purified linear vector from the glass with low salt buffer. These reagents are available in a commercial kit known as "Geneclean" (from Bio 101).

PCR products often are contaminated by additional DNAs that result from interactions between the primer and template at sites other than those intended. With luck, these unwanted products will be minor in quantity and have a different molecular weight than the desired DNA. If necessary, use agarose gel electrophoresis and the "Geneclean" procedure to isolate DNA of the proper size. We expect this step will not be necessary here.

Protocol

1. Prepare a 1% agarose gel (0.4 g agarose in 40 mL 1× TAE; heat in microwave until dissolved; add 1.5 μL of ethidium bromide; pour and insert comb).
2. To the restriction digests of pGEX2, add 4 μL 6× load buffer. Load the entire sample into 1 well. In separate lanes, load each single-cut control and uncut vector. Separate DNA by electrophoresis at 100 V (toward positive electrode).
3. Observe DNA under UV light. To protect DNA, use minimum fluence and long wavelength (365 nm). Set up a workstation with a long-wavelength UV light on a stand over a piece of clean glass. Place gel on glass and excise the portion containing the desired DNA band with a clean spatula and place it in a weighed 1.5-mL microfuge tube.
4. Weigh the agarose (in the tube).
5. Assume 1 g of agarose = 1 mL. Add 3 volumes of NaI solution (6 *M*).
6. Incubate at 55°C for 5 minutes, or until the agarose dissolves.

7. Add 5 μL of glass milk for solutions containing 5 μg or less of DNA. (Add 1 μL glass milk for each additional 1 μg of DNA.) Vortex.
8. **Important:** Incubate 5 to 10 minutes at room temperature to ensure that the DNA binds to the glass beads. Rock tube gently.
9. Centrifuge 5 seconds in microfuge.
10. Remove the supernatant with a Pipetman.
11. Wash (resuspend, centrifuge, decant supernatant, resuspend) the glass milk–DNA pellet 3 times with 0.5 mL (per wash) of NEW wash solution.
12. Resuspend the glass milk–DNA pellet in 10 μL of TE (10/0.1).
13. Incubate at 55°C for 5 minutes.
14. Centrifuge in microfuge for 30 seconds, then recover DNA contained in the supernatant.
15. To measure recovery, examine by gel electrophoresis (be sure you include the uncut plasmid on the gel for comparison), or determine the OD at 260 nm [OD_{260} nm × dilution factor × 50 = μg/mL DNA]. Recovery should average 80% of input.

Solutions for DNA Purification Using Glass Milk

Geneclean kit (Bio 101)

TAE:

0.04 *M* Tris–acetate, 0.002 *M* EDTA

1 L of 50×: 242 g Tris base
57.1 mL glacial acetic acid
100 mL 0.5 *M* EDTA pH 8.0

TE (10/0.1): 10 mM Tris, pH 7.5; + 0.1 mM EDTA

6× Load buffer: 0.05% bromphenol blue; 40% (w/v) glycerol in H_2O

To prepare your own reagents, if you do not want to use a commercial kit:

Glass milk solution: Silica 325 mesh: powdered flint glass from ceramic supply store; resuspend 250 mL powder in 750 mL water. Stir 1 hour then settle 1 hour. Take supernatant and repeat settling. Take second supernatant and centrifuge at 5000 rpm for 10 minutes in Sorvall GSA. Resuspend pellet in 200 mL water. Add nitric acid to 50%; steam for 4 to 5 hours. Wash 5 to 6 times (by centrifugation) with several volumes of water until supernatant is pH 7. Store at 4°C as a 50% slurry in water. Two hundred fifty milliliters of powder yields about 25 g of fines.

NaI solution (6.05 *M*): Mix 90.8 g NaI + 1.5 g Na_2SO_3 + water to 100 mL. Filter through Whatman #1 paper to remove undissolved sodium sulfite. Add another 0.5 g sodium sulfite; store in dark at 4°C.

Ethanol (NEW) wash: 50% ethanol + 0.1 *M* NaCl + 120 mM Tris, pH 7.5, + 1 mM EDTA

Notes

Notes

E. LIGATION OF PCR PRODUCT TO pGEX2 VECTOR

Introduction

DNA restriction fragments with compatible cohesive ends (or blunt ends) can be joined by phage T4 DNA ligase to form recombinant DNA molecules. Two competing reactions occur during ligation: intermolecular joining of 2 fragments at 1 end to produce a linear product, and intramolecular joining of opposite ends of a single molecule to form a circular DNA. Successful insertion of a restriction fragment into a plasmid vector requires both types of ligation. First, one end of the desired restriction fragment must ligate to one end of the plasmid vector (which has also been made linear by cleavage with an appropriate restriction endonuclease). Next, the free ends of this 2-component ligation product must join to form a circular recombinant plasmid (vector + insert) capable of transforming *E. coli*.

The total concentration of DNA termini in the ligation reaction is the most important factor controlling whether ligated DNAs will form primarily linear multimers or circular monomers. High concentrations favor linear multimers whereas low DNA concentrations favor circles. The likelihood that the ends of a single molecule will join to each other depends on its length: as their length increases, DNA molecules are less likely to form circles. The probability that the ends of a linear vector molecule will encounter the desired restriction fragment also depends on the relative abundance of each DNA and the number of other (competing) restriction fragments in the reaction. Thus, DNA concentration, fragment length, insert-to-vector ratio, and complexity of the DNA fragment population all affect the outcome of a ligation reaction.

Ligation requires ATP, which is usually included in the ligase buffer supplied by the manufacturer.

Protocol

1. Heat DNAs at 68°C for 15 minutes to inactivate restriction endonucleases (some other enzymes require phenol extraction followed by chloroform extraction and ethanol precipitation).
2. Mix: 100 ng of (linear) vector DNA
   500 ng of insert DNA (restricted PCR product; use one-fourth of the digest)
   2 μL of 10× ligase buffer
   H_2O to bring to 20 μL
3. As a control, prepare a ligation using 100 ng cut pGEX plasmid with no insert. This DNA should **not** transform *E. coli* because pGEX2 cut with both *Eco*RI and *Bam*HI **cannot** form circles.
4. Add 0.5 μL (200 units) of T4 DNA ligase to each.
5. Incubate at 15°C overnight.

Solutions for ligation

Use ligase buffer provided by the supplier, store frozen, and thaw on ice to protect the ATP.

New England Biolabs (NEB) T4 DNA ligase buffer: 50 mM Tris-HCl, pH 7.8, + 10 mM $MgCl_2$ + 1 mM ATP + 10 mM dithiothreitol + 25 μg/mL bovine serum albumin

Notes

Notes

F. TRANSFORMATION OF *E. COLI* WITH THE LIGATED PLASMID

Introduction

To introduce plasmid DNA into *E. coli,* grow the host cells to mid-log phase and incubate the bacteria (in the presence of plasmid DNA) in a buffer formulated to render the cell walls permeable to DNA. A heat shock completes the procedure. Such treatment stresses the bacteria; 30% survival (70% killing) is typical. This procedure does not work with linear DNA, which is sensitive to exonuclease attack. Different strains of *E. coli* vary greatly in transformation efficiency, and some strains (notably HB101) do not respond well to the RbCl method described here. For HB101, substitute 75 mM $CaCl_2$ for RbCl buffer. Note that "competent" cells (cells capable of taking up foreign DNA) can be prepared by this method and frozen for future use.

As host for the pGEX2 clones, you will use *E. coli* strain DH5α [F'/*supE44 hsdR17* (r_K- m_K+)*recA1 endA1 gyrA96* (NaI^r)*thi-1 relA1 Δ(lacIZYA− argF)U169 deoR* (Φ80*dlacΔ(lacZ)M15)*. DH5α transforms at a high efficiency.

Do the following transformations and controls:

a. Control 1: follow transformation procedure, but with no added DNA

b. Control 2: transform with uncut pGEX2 plasmid (use 10 ng)

c. Control 3: transform with cut pGEX2 (unligated; use 100 ng)

d. Control 4: transform with cut pGEX2 (ligated; use 100 ng)

e. Experimental: transform with cut, ligated pGEX2 and PCR DNA

Safety Precautions

You will be working with cultures of bacteria. Review aseptic techniques. Do not contaminate the buffers or media.

Protocol

Beforehand (TAs)

1. Inoculate 0.1 to 0.2 mL of fresh overnight culture into 10 mL of L broth. Grow cells in L broth to 1×10^8 cells/mL (a reading of 60 on a Klett colorimeter, or a spectrophotometer reading of 0.4 to 0.6 at 600 nm); prepare 5 mL for each transformation.

Students start here:

2. For 5 transformations, obtain 25 mL of cells (5 mL for each transformation). Centrifuge 5 minutes, 8000 rpm, 4°C in Sorvall SS-34 rotor. Use sterile 50-mL Nalgene tubes.
3. Discard supernatant. Resuspend pellet in 5 ml ice-cold RbCl buffer (sterile). Divide into 5 sterile 1.5-mL microfuge tubes; hold on ice 5 minutes.
4. Centrifuge 1 minute, 14,000 rpm, in microfuge.
5. Resuspend pellet in 0.2 mL ice-cold RbCl buffer.
6. Mix DNA with cells; use entire ligation, or controls as listed above. Hold on ice for 60 minutes.
7. Incubate in 42°C water bath for 2 minutes.
8. Add 1 mL of L broth (no antibiotics) to cells. Place in sterile test tube.
9. (optional) Incubate with aeration at 37°C for 45 minutes. Don't do this step if you want to avoid sib-

ling transformants, which are duplicate clones that arose from one transformed cell that divided during the 45-minute incubation. Because our goal in this experiment is to obtain clones, we do not need to avoid sibling transformants.

10. Pipet into microfuge tube. Pellet in microfuge; resuspend in 100 μL L broth.
11. Spread cells on selective agar plates.
12. Incubate at 37°C overnight.
13. Count colonies on each plate.
14. Use sterile toothpicks to pick single colonies and inoculate them into 2 ml of L broth (containing 100 μg/ml ampicillin). With the same toothpick, streak transformants on selective agar plates to purify single colonies from the transformants. You should always streak any transformants you intend to keep before storing them as permanent freezer stocks.

    Each team should pick 4 colonies: 1 from positive control 2 (uncut pGEX2) and 3 experimentals, which are putative recombinant clones.
15. Incubate the 2-mL cultures with aeration at 37°C overnight.

Solutions for Transformation

RbCl buffer: 100 mM RbCl + 45 mM $MnCl_2$ + 10 mM $CaCl_2$ + 35 mM KCH_3CO_2 + 15% sucrose, pH 6.0; filter sterilize

Mix 1.21 g RbCl + 0.89 g $MnCl_2 \cdot 4\ H_2O$ + 0.147 g $CaCl_2 \cdot 2\ H_2O$ + 0.344 g KCH_3CO_2 + 15 g sucrose in 100 mL total volume; adjust pH; filter through sterile 0.2-μm filter (Gelman HT-200) into sterile vacuum flask

L broth: 10 g/L tryptone (Difco) + 5 g/L yeast extract (Difco) + 10 g/L NaCl, pH 7.0 (autoclave 22 minutes)

L agar: add 15 g Difco agar to 1 L of L broth; autoclave

Selective agar plates: from 100× or 1000× stock, add sterile ampicillin to 100 μg/mL. Dissolve ampicillin in sterile water and store frozen at −20°C.

To store competent cells: proceed through step 5. **Do not** add transforming DNA. Incubate on ice for 60 minutes. Add an equal volume of ice-cold 30% glycerol. Freeze at −70°C. To use remove from freezer, add transforming DNA, incubate on ice for 30 to 60 minutes, and proceed with heat shock (step 7).

Notes

Notes

G. SMALL-SCALE PREPARATION OF PLASMID DNA BY THE ALKALINE LYSIS METHOD

Introduction

Use this procedure to purify small amounts of plasmid DNA for restriction endonuclease analysis. This rapid method allows us to prepare DNA from as many as 72 bacterial cultures at one time. The preparation will also contain RNA, some protein and cellular debris, and some sheared linear fragments of chromosomal DNA. Usually these impurities do not interfere with restriction analysis and many other procedures.

For many purposes, plasmid DNA is treated with RNase. You will not add RNase because the restriction fragments are larger than 1000 base pairs (bp), and the RNA will not interfere. Avoid RNase if future experiments will use RNA. Because your DNA preparations will contain RNA, you will not quantify them; RNA also absorbs at 260 nm.

DNA preparation kits are available from biotechnology companies. These kits produce DNA suitable for sequencing; however, the alkaline lysis method is much less expensive than a kit.

Protocol

1. Centrifuge 1.5 mL of overnight culture for 1 minute at 14,000 rpm in microfuge (use 1.5-mL tubes).
2. Discard supernatant and resuspend cells in 0.1 mL of TE (25/10).
3. Add 0.2 mL of alkaline lysis buffer (make fresh from NaOH and SDS). Mix **gently** (invert tube several times); excess vigor will increase contamination by sheared chromosomal DNA fragments.

4. Add 0.15 mL of 3 *M* potassium acetate buffer; mix **gently.**
5. Centrifuge 3 minutes in microfuge at full speed (14,000 rpm).
6. Remove supernatant to a 1.5-mL microfuge tube and mix with 1 mL 95% ethanol; hold 10 minutes at **room** temperature (or 4°C).
7. Centrifuge 3 minutes in microfuge.
8. Discard supernatant; dissolve pellet in 100 μL TE (10/0.1); add 50 μL of 7.5 *M* ammonium acetate.
9. Precipitate with 300 μL of 95% ethanol (on ice, 10 minutes).
10. Centrifuge 3 minutes in microfuge; remove all ethanol. Wash the pellet with 70% ethanol: add 0.5 mL ethanol to the tube, invert **gently,** then remove ethanol. Air dry pellet to remove traces of ethanol.
11. Dissolve pellet in 60 μL of TE (10/0.1). Use 2 to 5 μL per sample for restriction analysis and agarose gel electrophoresis; treat with 1 μL RNase **(not needed here)** to examine DNA restriction fragments smaller than 1000 bp.

Note: Many protocols omit the 2nd ethanol precipitation. Follow the procedure through step 7. Discard supernatant. Wash the pellet with 70% ethanol and air dry. Dissolve the pellet in 60 μL TE.

Solutions for Plasmid DNA Preparation (Alkaline Lysis)

TE (25/10): 25 mM Tris, pH 7.5; 10 mM EDTA (autoclave)

SDS–NaOH lysis solution (make fresh before use):
200 mM NaOH; 1% SDS

Mix 0.67 mL of 3 *M* NaOH + 0.4 mL of 25% SDS + 9 mL H_2O
(3 *M* NaOH = 12 g/100 mL)

3 *M* potassium acetate solution: 3 *M* K+ 5 *M* CH_3CO_2- (autoclave)

Mix 60 mL of 5 *M* potassium acetate (29.45 g/60 mL) + 11.5 mL of acetic acid + 28.5 mL of H_2O

Ammonium acetate solution: 7.5 *M* ammonium acetate = 57.8 g/100 mL (autoclave)

TE (10/0.1): 10 mM Tris, pH 7.5; 0.1 mM EDTA (autoclave)

95% Ethanol

Notes

H. RESTRICTION ANALYSIS

Introduction

Analysis of recombinant plasmids (created by restriction and ligation in vitro) to determine whether cloning was successful involves: 1) digestion of recombinant plasmid DNAs with appropriate restriction endonucleases; 2) separation of restriction fragments according to their size via electrophoresis through an agarose gel; and 3) estimation of restriction fragment molecular weights.

Usually we perform more than one restriction analysis to confirm the structure of a particular recombinant plasmid. For example, we often choose to excise the insert from the vector with the same enzymes used to make the recombinant plasmid. This will show whether the putative recombinant plasmid contains an inserted restriction fragment of the proper size. We can also map a restriction site found inside the inserted fragment relative to a site in the vector to determine the orientation of the insert in the vector. This provides independent evidence that the recombinant plasmid has the expected structure.

The pGEX2–*virD2* plasmid you constructed should contain a single *Eco*RI site at one junction between the PCR-produced 1280-bp DNA fragment and the 4948-bp vector. We expect digestion with *Eco*RI to produce a single fragment of 6218 bp. The other vector–insert junction is a hybrid *Bam*HI/*Bgl*II site, which cannot be cleaved by either enzyme. The insert fragment contains a single *Pst*I site 121 bp from its *Bgl*II end and 1159 bp from its *Eco*RI end. The vector contains a single *Pst*I site 957 bp from its *Eco*RI site and 3981 bp from its *Bam*HI site. Thus, *Pst*I digestion of the plasmid we constructed should produce fragments of 2116 and 4102 bp.

Technical Tip

Keep enzymes chilled (store at −20°C in non-frost-free freezer).

Protocol

1. Mix **2** of the following reactions, one for *Eco*RI and one for *Pst*I:

   Mix:
   5 μL of miniprep DNA (200 to 800 ng)
   1.5 μL of 10× reaction buffer (supplied by manufacturer) (NEBuffer *Eco*RI for *Eco*RI; NEBuffer 3 for *Pst*I);

   Bring to 15 μL with water (8.5 μL).

2. Add 2 to 5 units of restriction endonuclease (usually about 0.5 μL)
3. Incubate at proper temperature (37°C for most enzymes) for 1 or more hours.
4. Mix agarose with 1× TAE buffer (0.8% = 0.2 g/25 mL for minigel); melt in microwave; add 1.25 μL of 10 mg/mL ethidium bromide. Pour molten agarose into bed with comb (well-former) in place; allow to solidify; cover with buffer and remove comb.
5. Place 5 μL of each restriction digest in fresh tubes. Add 1 μL 6× load buffer to each and load in individual wells. Note: one-third of your restriction digest should form visible bands. However, if your DNA yield was low, the bands may appear faint; you can use the remaining two-thirds to repeat the electrophoresis.
6. Fill buffer chambers with 1× TAE; apply 100 V; DNA will migrate to the positive electrode.

7. When bromphenol blue dye reaches bottom of gel, observe DNA under UV light; photograph.

Solutions for Restriction Analysis

TAE: 0.04 *M* Tris–acetate, 0.002 *M* EDTA
1 L of 50×: 242 g Tris base
57.1 mL glacial acetic acid
100 mL 0.5 *M* EDTA pH 8.0

NEBuffer *Eco*RI: 50 mM NaCl, 100 mM Tris–HCl, 10 mM $MgCl_2$, 0.025% Triton X-100 (pH 7.5 at 25°C)

NEBuffer 3: 50 mM Tris–HCl, 10 mM $MgCl_2$, 100 mM NaCl, 1 mM dithiothreitol (pH 7.9 at 25°C)

6× Load buffer: 0.05% bromphenol blue; 40% glycerol in water

Notes

EXERCISE 1. Study Questions

Cesium Chloride Gradient

1. Sometimes undigested plasmid DNA preparations isolated from CsCl gradients show 4 bands (in the upper half of the gel) on agarose gel electrophoresis. What form of plasmid DNA is in each band? How could you test whether the plasmid DNA in a particular band is supercoiled (also called covalently closed circular), relaxed circular (also called nicked circular or open circular), or linear?

2. Why does each DNA form migrate differently during agarose gel electrophoresis?

3. What is the extremely bright broad band of material near the bottom of the gel? How does its presence affect your estimate of the plasmid DNA concentration based on OD_{260} readings? What would you do to eliminate this problem?

PCR

1. What does each control in the PCR reaction tell you? What other controls might be useful?

2. Several parameters affect the melting temperature (T_m) of DNA:

   a. base composition

   b. salt concentration

   c. formamide: 1% formamide lowers T_m by 0.65°C

   d. DNA sequence similarity: every 1% decrease in similarity lowers T_m by 1.5°C

   e. length of DNA: T_m (short duplex) = T_m (long duplex) − (500 ÷ bp of short duplex)

   The following equation is used to estimate the melting point of duplex DNA.

   $$T_m = 16.6 \log [Na^+] + 0.41 (\% G+C) + 81.5$$

   Does an increase in salt concentration raise or lower the T_m of DNA? Why?

3. a. Look at your gel analysis of the PCR products. Is the brightest band in each lane the desired prod-

uct? How can you tell? What else could you do to confirm the identity of the PCR product? Is such confirmation important?

b. Estimate the molecular weight of the major PCR product; do a semilog plot of molecular weight versus migration distance for the 1-kb ladder marker.

c. Did the kit primer and template give a PCR product of the anticipated molecular weight?

d. On some gels, the no-template control produced a faint band the same size as the desired PCR product. Explain what probably led to the appearance of this band and what steps you could take to prevent this problem in the future.

e. What are the other bands? How might you prevent synthesis of these minor PCR products?

Cloning

1. During the cloning you cut pGEX2 (4948 bp) with *Eco*RI and *Bam*HI (which cut next to each other), and you attempted to insert your PCR product (1280 bp), after cleavage with *Eco*RI and *Bgl*II, into the compatible cohesive ends of the cut pGEX2. Why are the ends of these DNAs compatible even though they are cut with different restriction endonucleases?

2. Why did you heat the restriction digest at 65°C for 10 minutes before you added T4 DNA ligase?

3. Which of the enzymes you used can be inactivated by heat (65°C) and which cannot (see the appendix in the New England Biolabs [NEB] catalog)? Does it matter? Explain.

4. Some restriction endonucleases do not cut well when the target site is at or near the end of a DNA molecule, such as the PCR product you used. Does this observation apply to your cloning experiment? What is the likely consequence if one of the enzymes failed to cut its target in the PCR product?

5. Why did we design a *Bgl*II site into the PCR product instead of a *Bam*HI site? (Use your imagination.)

6. What is the molar ratio between vector and insert in the ligation you did?

7. What is the purpose of each control in the transformation procedure?

8. What was your transformation efficiency? How does the efficiency of your cells compare with commercial competent cells?

9. Make a drawing of the recombinant plasmids we made. Include restriction enzyme recognition sites, and label the restriction fragments with their lengths in base pairs.

10. Look at the gel of the restriction analysis of plasmid DNAs isolated from transformants in your cloning experiment. Measure the distance each band has migrated. Use the known molecular weight standards on your gel to plot a molecular weight versus migration distance curve, and use this curve to estimate the molecular weight of each unknown restriction fragment. Do any of the lanes contain the desired clones? Can you explain what may have happened in the lanes that do not?

Notes

EXERCISE

2

Protein Expression, Purification, and Analysis

Background

The plasmid pGEX2–VirD2 (the PCR product *virD2* inserted into pGEX2) will express the fusion protein Gst–VirD2 (Gst = glutathione-*S*-transferase). Transcription of this hybrid gene will come from the *tac* promoter located upstream of *gst*.

The *tac* promoter is a hybrid of the *E. coli trp* and *lac* promoters; it is a very strong promoter regulated in the same manner as the *lac* promoter. The *lacI* gene encodes the Lac repressor, which binds to *lac* promoter/operator regions and inhibits transcription. In *lacI*+ cells, a single copy *lac* (or *tac*) promoter will be repressed by *lacI* unless an inducer, such as lactose or IPTG (isopropylthiogalactoside), is present. In multicopy plasmids (for example, pUC or pGEX), multiple copies of the *tac* promoter titrate all of the Lac repressor made in *lacI*+ strains. To regulate *tac* promoters on these multicopy plasmids, cells must overproduce the LacI protein. Strains carrying the mutant allele *lacI*^{q} overexpress LacI at levels sufficient to regulate the *tac* promoter on multicopy plasmids. Full repression of the gene is not necessary, because the Gst–VirD2 fusion protein is not deleterious to *E. coli.*

When expressing foreign proteins in *E. coli,* protein insolubility frequently presents a significant problem. If insolubility problems arise, temperature, inducer concentration, and promoter strength may all be varied to increase the percentage of soluble recombinant protein. Fusion to Gst typically increases solubility.

Protease degradation of foreign protein in wild-type *E. coli* strains can also present problems. Various protease-deficient strains are sometimes used to combat this. For example, *E. coli* strain PR78 is deficient in the *lon* protease. PR78 contains Tn*10* (a transposable element that confers tetracycline resistance) inserted into the *lon* gene (*lon*::Tn*10*) to inactivate its expression. Lon− strains often increase the stability of foreign proteins.

After expression of Gst–VirD2, you will prepare crude cell extracts and purify the fusion protein by glutathione–Sepharose affinity chromatography. The glutathione-*S*-transferase moiety of the fusion protein binds to glutathione. After Gst–VirD2 is mixed with the glutathione–Sepharose beads and permitted to adhere, washing removes most other proteins and cellular debris. We will analyze the crude and purified extracts by SDS-polyacrylamide gel electrophoresis. We will detect the proteins in the gel chemically by the silver stain technique and immunologically by the Western blot technique.

Steps of the experiment are

A. Express and purify fusion protein

B. SDS-polyacrylamide gel electrophoresis

C. Silver stain detection of proteins

D. Western blot (immunoblot) detection of fusion protein

Notes

A. EXPRESSION AND PURIFICATION OF A FUSION PROTEIN

Introduction

You will express Gst–VirD2 in the *E. coli* strain MC1061. To help prevent protein degradation during protein purification, we will include 3 protease inhibitors in our buffer solutions and perform purification steps at 0 to 4°C.

You will extract, purify, and analyze proteins from 3 strains:

1. MC1061 (pGEX2–*virD2*)
2. MC1061 (pGEX2)
3. MC1061

The strain containing pGEX2 will express wild-type Gst protein, which will assure that the glutathione–Sepharose affinity purification works properly. Plasmidless MC1061 will show which chromosomally encoded proteins are not removed by the purification procedure.

Protease degradation can be a significant problem when a protein is expressed in wild-type *E. coli* strains. To examine this, a control group will express our protein in *E. coli* strain PR78, deficient in the Lon protease. Our experience shows that the Gst–VirD2 fusion protein is significantly more stable in Lon− than in Lon+ strains.

The control group will extract, purify, and analyze proteins from these strains:

1. PR78 (pGEX2–*virD2*)
2. PR78 (pGEX2)
3. PR78
4. MC1061 (pGEX2–*virD2*)—compare with PR78

Because MC1061 is not Lon protease deficient, you will see what effect this protease has on yield of recombinant protein.

Safety Precaution

Phenylmethylsulfonyl fluoride (PMSF) is dangerous if inhaled, swallowed, or absorbed through skin. Avoid contact. PMSF is inactivated in aqueous solution. Therefore, prepare a 10× stock solution in isopropanol (7.5 mM = 1.3 mg/mL) and freeze in aliquots. Add one-tenth volume of PMSF stock to PBS just before use. Aqueous solutions of PMSF can be discarded after raising the pH to greater than 8.6 and holding at room temperature for several hours.

Protocol

1. (TAs) Inoculate 2 mL of L broth (+50 μg/mL ampicillin) with 40 μL of fresh overnight culture of MC1061 harboring pGEX2–VirD2 and MC1061 harboring pGEX2. Inoculate 2 mL of LB (no antibiotic) with MC1061. Repeat for PR78. Incubate all cultures at 37°C with aeration for 2.5 to 3 hours until the culture is visibly turbid (Klett = 60). Chill microfuges in the 4°C cooler.
2. **(Students start here)** Add 20 μL of 100 mM IPTG (final concentration 1 mM); continue incubation for 1 to 2 hours.
3. Just before use, prepare a fresh solution of ice-cold PBS + protease inhibitors.
4. Centrifuge cells for 1 minute in microfuge, discard supernatant, and resuspend cells in 300 μL of ice-cold PBS + protease inhibitors. Remove and freeze (at −20°C) two 5-μL samples of intact cells for gels.

5. Lyse the remainder by two 10-second sonicator pulses. From this point, **keep all samples on ice** and process as rapidly as possible.
6. Centrifuge lysed cells for 5 minutes at 4°C in microfuge; retain supernatant.
7. Add 50 μL of glutathione–Sepharose–PBS slurry to supernatant (300 μL); mix gently at room temperature for 2 minutes.
8. Add 1 mL of PBS–protease inhibitors, vortex 30 seconds, centrifuge 5 seconds in microfuge, and remove supernatant.
9. Repeat step 8 three times; allow to stand for 5 minutes between vortex and centrifugation.
10. Remove two 10-μL samples of washed beads (pellet) for analysis by SDS-polyacrylamide gel electrophoresis (SDS-PAGE). To the 10-μL samples of beads, add an equal volume of 2× sample buffer and boil for 1 minute.
11. Microfuge 15 seconds, then hold samples on ice until ready to load the SDS-polyacrylamide gel.

Solutions for Protein Expression and Purification

100 mM IPTG: dissolve in sterile distilled water in a sterile tube; use a sterile spatula; store at −20°C.

Phosphate-buffered saline (PBS):

To make 1 L add	Final concentrations are
8.0 g NaCl	137 mM NaCl
0.2 g KCl	2.7 mM KCl
1.15 g $Na_2HPO_4 \cdot 7\ H_2O$	4.3 mM $Na_2HPO_4 \cdot 7\ H_2O$
0.2 g KH_2PO_4	1.4 mM KH_2PO_4

Protease inhibitor stocks:

7.5 mM PMSF; dissolve in isopropanol
100 mM Phenanthroline; dissolve in methanol
0.5 *M* EDTA, pH 8.0
2-Mercaptoethanol (14.4 *M*)
10 mg/mL Aprotinin

For each 10 mL of PBS add	Final concentrations are
1 mL PMSF	0.75 mM PMSF
500 μL phenanthroline	5 mM phenanthroline
200 μL EDTA	10 mM EDTA
6 μL mercaptoethanol	10 mM mercaptoethanol
2 μL aprotinin	2 μg/mL aprotinin

(A number of other protease inhibitors may also be used.)

Glutathione–Sepharose (agarose) slurry: Wash 1 mL of Pharmacia glutathione–Sepharose or Sigma glutathione agarose slurry 3 times with excess PBS; store as a 50% slurry in PBS at 4°C.

2× Sample buffer:
10 mL of 0.5M Tris, pH8.0
10 mL of 10% SDS
5 mL of 2-mercaptoethanol
10 mL glycerol
65.4 mg EDTA
4.0 mg bromphenol blue
65 mL H_2O

Notes

Notes

B. SDS-POLYACRYLAMIDE GEL ELECTROPHORESIS

Introduction

We use SDS-PAGE to separate proteins primarily according to their size. The anionic detergent SDS (sodium dodecyl sulfate) binds to the proteins, causes the polypeptides to denature, and imparts a large negative charge on the proteins. This SDS charge masks any charge normally present and equalizes the charge along the protein molecules. When an electric current is applied during SDS-PAGE, each protein in a sample migrates toward the anode at a rate inversely proportional to its molecular weight. For highly positively charged proteins, molecular weights calculated from migration during SDS-PAGE may differ significantly from the true molecular weight.

Polyacrylamide gel polymerization is catalyzed by tetramethyl ethylenediamine (TEMED) and ammonium persulfate. Once these reagents are added, the gel will begin to polymerize. Oxygen inhibits polymerization. The butanol overlay protects the gel surface from oxygen. You can tell when the gel is fully polymerized because a well-defined interface will form between the gel and the overlay; this usually takes about 15 minutes.

You will use a discontinuous gel. The purpose of this gel system is to concentrate the protein sample into a small volume, which increases the resolution of the gel. The top component, called the stacking gel, will be 6% acrylamide, pH 6.8. The sample buffer is also pH 6.8. The bottom component, called the resolving gel, is 12% acrylamide, pH 8.8. The sample moves quickly through the stacking gel and forms a tight band as it enters the resolving gel. The resolving gel then separates proteins according to their molecular weights.

Proteins are usually detected after SDS-PAGE by staining with silver or Coomassie blue (R250), or by the Western blot procedure. Proteins may be radiolabeled during expression (usually with ^{35}S-methionine) and detected by autoradiography. Note that all of these detection methods are qualitative; do not use SDS-PAGE to quantify the amount of protein present in a sample.

Safety Precautions

Acrylamide is a neurotoxin; use gloves. Acrylamide is available in solution form; this prevents dust inhalation during handling.

Protocol

1. **Clean** and assemble glass plates and spacers. Clamp in place.
2. Prepare and pour resolving gel

   Combine: 3 mL of the 40% acrylamide stock
   2.5 mL of 1.5 *M* Tris, pH 8.8
   100 μL of 10% SDS
   4.3 mL of water
   in a flask and swirl gently to mix.

   Add: 100 μL of 10% ammonium persulfate
   10 μL of TEMED
   to mixture and swirl gently to mix.

   Pour gel using automatic pipettor (Pipet-Aid) or Pasteur pipet, and overlay with water-saturated *n*-butanol using a Pasteur pipet.
3. After resolving gel has polymerized, remove butanol, wash top surface with distilled water, dry gel surface with a Kimwipe, and place comb in position.
4. Prepare and pour stacking gel

Combine: 0.75 mL of the 40% acrylamide stock
1.25 mL of 0.5 *M* Tris, pH 6.8
50 μL of 10% SDS
2.93 mL of water
in a flask and swirl gently to mix.

Add: 50 μL of 10% ammonium persulfate
5 μL of TEMED
to mixture, swirl gently to mix, and pour gel using automatic pipettor or Pasteur pipet. To avoid catching air bubbles underneath the gel comb, slowly pipet gel solution down the inside edge of the apparatus.

5. After stacking gel has polymerized, pipet water onto top of gel, remove comb, and rinse wells with distilled water. Rinsing removes unpolymerized gel material; failure to rinse leads to uneven sample wells.

6. Remove bottom spacer, place gel in apparatus, add electrophoresis buffer, and remove air pocket underneath gel with a buffer-filled syringe and bent needle.

7. Add 1 volume of 2× sample buffer to each sample (except Sepharose bead samples, which are already prepared) and boil for 1 minute. For molecular weight marker, use 7.5 μL marker with 7.5 μL of sample buffer.

8. Load samples. Secure electrodes and apply 100 V (constant voltage) while samples migrate through stacking gel. When samples reach resolving gel, increase to 200 V. Proteins will migrate toward the positive electrode (anode). Continue electrophoresis until the bromphenol blue dye reaches the bottom of the gel. This will take approximately 20 minutes.

SDS-POLYACRYLAMIDE GELS: LOAD ORDER

A. MC1061

Lane	Plasmid	Strain	Sample	Volume Sample (μL)	Volume 2× SB (μL)
1			molecular weight marker	7.5	7.5
2	none	MC1061	resuspended cells	5	5
3	pGEX2	MC1061	resuspended cells	5	5
4	pGEX2–VirD2	MC1061	resuspended cells	5	5
5			molecular weight marker	7.5	7.5
6	none	MC1061	purified extract	10	10
7	pGEX2	MC1061	purified extract	10	10
8	pGEX2–VirD2	MC1061	purified extract	10	10
9			molecular weight marker	7.5	7.5

B. PR78 (*lon*−) controls

Lane	Plasmid	Strain	Sample	Volume Sample (μL)	Volume 2× SB (μL)
1			molecular weight marker	7.5	7.5
2	none	PR78	resuspended cells	5	5
3	pGEX2	PR78	resuspended cells	5	5
4	pGEX2–VirD2	PR78	resuspended cells	5	5
5	pGEX2–VirD2	MC1061	resuspended cells	5	5
6	none	PR78	purified extract	10	10
7	pGEX2	PR78	purified extract	10	10
8	pGEX2–VirD2	PR78	purified extract	10	10
9	pGEX2–VirD2	MC1061	purified extract	10	10
10			molecular weight marker	7.5	7.5

Solutions for SDS-Polyacrylamide Gel Electrophoresis

Acrylamide stock: use purchased 40% acrylamide, 29:1 acrylamide:bis-acrylamide. Or use 40% acrylamide stock:

Combine 387 g of acrylamide + 13 g bis-acrylamide (bis controls the extent of cross-linking). Bring to 1 L with distilled water.

0.5 *M* Tris, pH 6.8 (use Tris base)

1.5 *M* Tris, pH 8.8 (use Tris base)

10% SDS

TEMED

10% Ammonium persulfate: make fresh and store at 4°C for no longer than 1 week; use distilled water.

TG-SDS running buffer:	Final concentration
57.6 g glycine	380 mM
12.0 g Tris base	50 mM
2.0 g SDS	0.1 % (w/v)
Bring to 2 L with water	

2× Sample buffer:

10 mL of 0.5 *M* Tris, pH 6.8
10 mL of 10% SDS
5 mL of 2-mercaptoethanol
10 mL glycerol
65.4 mg EDTA
4.0 mg bromphenol blue
65 mL water

100 mL total volume

Molecular weight markers:

Bio-Rad Low Range Silver Stain SDS-PAGE standards
Bio-Rad Kaleidoscope Prestained Standards (for Western blot)

Notes

C. SILVER STAIN DETECTION OF PROTEINS

Introduction

We will use the ICN Rapid-Ag silver stain kit. If you have questions about the kit, refer to the instructions provided with the kit. Procedures are also available to prepare silver stain reagents from laboratory stocks. To prevent contamination of stock solutions, wear gloves while preparing solutions and use a fresh pipet for each measurement.

Technical Tip

Silver stain will detect fingerprints, glove prints, dust, and anything else that touches your polyacrylamide gel. Handle gels with gloves, and only at the edges.

Procedure

1. Prepare 200 mL of gel fixing solution (40% methanol + 10% glacial acetic acid).
2. **Do not handle gel.** Use a glass plate to support the gel; place both in a tray containing 200 mL of gel fixing solution and gently float the gel free of the glass plate. Soak the gel for at least 30 minutes with gentle agitation (at room temperature). There is no time limit on this step; it can continue overnight.
3. Rinse with 200 mL water, then agitate in 200 mL water for 15 minutes. Aspirate to remove solutions; this reduces stress on the gel.
4. Prepare pretreat solution, silver stain, developer, and stop bath.

5. Remove water, add 200 mL pretreat solution, and agitate for 10 minutes.
6. Remove pretreat solution, add 200 mL silver stain solution, and agitate 30 minutes.
7. Remove silver stain and rinse gel with 200 mL water.
8. Wash 4 times with 200 mL water (each) for 2 minutes per wash.
9. Add 200 mL developer and agitate until bands reach the desired darkness. Stop development by adding 2 mL of glacial acetic acid.
10. Remove developer and soak gel in 200 mL of 1% glacial acetic acid for 5 minutes.
11. Remove dilute glacial acetic acid and soak gel in 200 mL water for 5 minutes.
12. Glycerol fix: soak gels in 20% ethanol, 10% glycerol for 30 minutes before drying. This prevents cracking of the gel.
13. Soak 2 pieces of cellophane in glycerol fix. Lay 1 piece of cellophane on plastic wrap. Place your gel on top. Pour 5 to 10 mL glycerol fix over the gel. With a rolling motion, lay the 2nd piece of cellophane over the gel. Remove bubbles. Lay the gel and cellophane on the plastic square, put the plastic frame on top, and clamp. (The apparatus resembles a picture frame and back, minus the glass; it holds the gel flat.) Dry for 2 days.

Solutions for Silver Stain Detection of Proteins

ICN Rapid-Ag Stain Kit

Gel fixing solution:	80 mL methanol
	20 mL glacial acetic acid
	100 mL water
	200 mL total
Pretreat solution:	
combine (sequentially):	80 mL methanol
	20 mL ethanol
	90 mL water
	10 mL ICN solution #1
swirl to mix	200 mL total
Silver stain:	
combine (sequentially):	10 mL ICN solution #2
	10 mL ICN solution #3
(swirl to dissipate	10 mL ICN solution #4
brown color)	170 mL water
	200 mL total
Developer:	380 mL water
	10 mL ICN #5
	10 mL ICN #6
	400 mL total

Stop bath: 1% glacial acetic acid

20% Ethanol, 10% glycerol

Notes

D. WESTERN BLOT (IMMUNOBLOT) DETECTION OF PROTEINS

Introduction

The Western blot procedure is used to detect a specific protein among a mixture of proteins, or to show that a particular purified (or partially purified) protein is recognized by a specific antiserum. After SDS-PAGE, proteins are transferred by electrophoresis (also called electroblotting) to a nitrocellulose, nylon, or PVDF (polyvinylidene difluoride) membrane. A blocking agent binds nonspecific sites on the membrane. Primary antibodies raised against an antigen (the protein of interest) then bind specifically to protein antigens fixed to the membrane. The primary antibody is bound by a secondary antibody, which may be conjugated to a number of different "reporter" enzymes or molecules such as alkaline phosphatase, peroxidase, or biotin. In our experiment, the secondary antibody is conjugated to alkaline phosphatase; alkaline phosphatase activity is detected by conversion of a colorless substrate, 5-bromo-4-chloro-3-indolyl phosphate (BCIP), into a blue indigo derivative.

Because the strength of the Western blot signal (color reaction) depends on both the affinity of the antibody for the antigen and the concentration of antigen present, the Western blot technique is **not** considered a quantitative measure of antigen concentration. An enzyme-linked immunosorbent assay (ELISA) **can** be used as a quantitative measure of antigen concentration. ELISAs are usually based on the same types of color reactions as immunoblots.

Two different types of antibody preparations are used for immunoblotting: polyclonal and monoclonal antibodies. Polyclonal antibodies are prepared by direct immunization of an animal with the antigen, and consist of the full repertoire of the animal's circulating antibodies. Polyclonal antisera may contain antibodies highly specific for your antigen and others that recognize spurious antigens. Monoclonal antibodies are prepared by repeated direct immunizations of an animal followed by dissection of the animal's spleen (source of antibody-producing cells) and fusion of those spleen cells with a tumor cell line. The resulting monoclonal cell line expresses antibodies to a single antigenic epitope.

Technical Tips

Nitrocellulose membranes are fragile and sensitive to fingerprints. **Pick up the membrane (in a corner) with clean forceps only.**

For experiments that use nylon or nitrocellulose membranes (Westerns, Southerns, Northerns), contaminants on the membrane cause significant background. Follow the times and temperatures for membrane washes in each of these procedures; skimping results in dirty blots.

Protocol

1. After SDS-PAGE, place the unstained gel in 50 mL standard transfer buffer and rock for 10 minutes.

2. Cut the nitrocellulose filter to the size of the gel, and cut 2 pieces of Whatman 3 MM paper to the size (15 × 20 cm) of the Scotchbrite pads supplied with the electroblot apparatus (Mini-Genie Immunoblotters from Idea Scientific). Soak the pads, nitrocellulose, and Whatman 3 MM paper in transfer buffer. Do not touch the gel or any membranes or filters with ungloved hands.

3. Lay down 1 pad and cover with a sheet of Whatman 3 MM. Place the gel on the 3 MM paper and lay the

nitrocellulose membrane over the gel. Lay a piece of Whatman 3 MM paper over the membrane. Use a pipet to roll out air bubbles between the gel and the membrane, and finally cover the entire sandwich with the other Scotchbrite pad.

4. Place the sandwich in the apparatus; it is extremely important that the sandwich is tight. Use additional Scotchbrite pads if necessary. Fill the chamber with transfer buffer, and apply 500 mA of current for approximately 30 minutes (this depends on the size of the protein you wish to transfer).

5. Wash the nitrocellulose membrane with rocking for 5 minutes in 15 mL distilled water (in a small plastic container).

6. Soak the membrane in 15 mL TBS for 10 minutes with rocking.

7. Soak membrane for 1 hour in 15 mL blocking solution (TBS + Tween 20 + gelatin), with rocking.

8. Pour off blocking solution. Wash the membrane in 15 mL TBS for 5 minutes with rocking.

9. Add 15 mL TBS. Add 15 μL of serum containing primary antibody (polyclonal antiserum raised in rabbit against glutathione-*S*-transferase antibody, 1 mg/mL); final concentration is 1 μg/mL in TBS. Mix thoroughly. Incubate, with rocking, at room temperature for 1 hour. This incubation can continue overnight.

10. Pour off buffer. Wash 3 times with 15 mL of TTBS for 5 minutes each with rocking.

11. Pour off buffer and add 15 mL TBS to dish. Now add 15 μL of secondary antibody (antiserum raised in goats against rabbit antibodies; the secondary antibody is conjugated to alkaline phosphatase). Incubate for 30 minutes at room temperature with rocking.

12. Pour off buffer, then wash 3 times in 15 mL TTBS for 5 minutes each.
13. Pour off buffer and replace with 15 mL TBS. Leave in TBS until ready to add substrate.
14. Pour off last wash, then add 10 mL of substrate solution. Incubate until color reaction is sufficient.
15. Rinse blot in distilled water. Dry on Whatman paper.

Solutions for Western Blot

Transfer buffer: 25 mM Tris (3 g/L Tris base)
192 mM glycine (14.4 g/L)
20% methanol (200 mL/L)
0.1% SDS (1 mL/L)

TBS (Tris-buffered saline): 20 mM Tris, pH 7.5
150 mM NaCl

TTBS: TBS + 0.1% Tween 20

Blocking solution: TBS
0.1% Tween 20
1% gelatin

Primary antibody: Upstate Biotechnologies anti-Gst; purified polyclonal IgG antibody raised in rabbits against Gst expressed by pGEX2.

ICN Western Blot Staining Kit: this kit supplies secondary antibody (goat polyclonal antiserum made against rabbit antigens), and color staining reagents (nitro blue tetrazolium and BCIP).

Notes

Notes

EXERCISE 2. Study Questions

1. Crude extracts of plasmidless MC1061 contained a number of proteins that reacted with the anti-Gst antibody. Do these bands have anything to do with Gst or the Gst–VirD2 fusion protein? Explain.

2. You saw bands on the Western blot that were plasmid-specific. We suspect that these bands correspond to Gst and Gst–VirD2. Measure the mobility of the molecular weight standards, plot a standard curve on semilog paper, and estimate the molecular weight of each plasmid-specific band. Can you conclude that these antibody-stained bands represent Gst and Gst–VirD2?

3. Do the antibody-stained bands on the Western blot match silver-stained bands on the previous gel?

4. There are 2 faint plasmid-specific bands corresponding to low molecular weight positions on some of the Western blots. How are these formed?

5. Must plasmid-specific bands correspond to proteins encoded by genes on the plasmid? How could you prove that a protein is plasmid encoded, and not simply regulated by a gene on a plasmid?

EXERCISE 3

Oligonucleotide-Directed Mutagenesis

Background

The ability to create specific mutations is one of the most powerful techniques for analysis of well-defined genes. With this technique, we can create single-base changes or small insertions and deletions. For example, we can introduce or destroy a restriction site, alter a promoter, or change the coding sequence of a gene.

Oligonucleotide-directed mutagenesis allows us to alter DNA in a specific location by hybridizing a complementary oligonucleotide primer containing the desired mutation to a circular single-stranded vector containing the target sequence. DNA polymerase extends the primer in vitro; DNA ligase covalently joins the ends of the newly extended strand. This heteroduplex DNA now contains the mutation in its newly synthesized strand and the original sequence in the template strand. On transformation into *E. coli,* DNA repair and replication resolve this mismatch to produce colonies with either mutant or wild-type target DNA.

Several different methods can increase the chances of recovering mutant-containing clones. You will use uracil-containing template prepared from a mutant *E. coli* strain. You will provide dTTP for in vitro DNA synthesis, so that the mutant strand will contain thymine, not uracil. The uracil-containing template strand will be selectively destroyed when transformed into a wild-type host strain, leaving behind the mutant strand.

In this experiment, you will delete 58 bp from *virD2* and replace them with a 7-bp sequence that includes an *Nru*I site. You begin by inserting a portion of *virD2* into the phagemid vector pUC119 (*virD2* + pUC119 = pMCB525). A phagemid vector is a plasmid that contains both a plasmid (ColE1) and single-strand DNA phage (M13) origins of replication. When cells that harbor a phagemid are infected with an M13 helper phage (M13KO7), the cells produce phage particles that contain single-stranded DNA composed of either phagemid DNA or helper phage DNA. Because M13KO7 replicates poorly in the presence of a phagemid (the helper phage has a slightly defective *ori*), most capsids contain phagemid DNA.

Experience shows that large restriction fragments inserted into phagemid vectors suffer rearrangements at a high frequency. Therefore, we will use a small restriction fragment, a 354-bp portion of *virD2* contained on a *Bam*HI–*Hin*dIII restriction fragment. To confirm the presence of the mutation, you will use these same restriction sites to excise the mutant portion of *virD2* from putative mutant derivatives of pMCB525. First you must insert this fragment into the corresponding sites of pUC119 to create the phagemid pMCB525. You will transform *E. coli* with pMCB525. The phagemid provides the host with antibiotic resistance, allowing you to select for transformed cells; when the host is plated on antibiotic-containing medium, only cells that contain the phagemid can grow. In addition, ligation of the insert into the phagemid disrupts *lacZ* present in the phagemid, allowing you to identify transformed cells that contain the insert.

You will prepare template for in vitro mutagenesis from a *dut– ung– E. coli* strain (CJ236) harboring the recombinant phagemid pMCB525. The *dut* mutation inactivates dUTPase, thereby increasing the cellular dUTP pool, which results in increased incorporation of uracil (in place of thymine) during DNA synthesis. The *ung* mutation inactivates uracil *N*-glycosylase, a DNA repair enzyme that removes uracil bases; this mutation prevents removal of uracil from the DNA. The uracil-containing phagemid DNA will serve as the template for in vitro DNA synthesis primed by the mutagenic oligonucleotide. The 5' end of this primer is the exact complement of a 14-base sequence in the template, and the 3' end is the complement of a separate 19-base region of the template. These 2 regions are separated by 58 bases in the template, but only 7 bases (which include an *Nru*I site) in the primer. Thus, successful mutagenesis will delete 58 bp from *virD2* and replace them with a 7-bp sequence that includes an *Nru*I site. You will screen putative mutants for a net loss of 51 bp from the *Bam*HI–*Hin*dIII fragment.

On completion of DNA synthesis, you will transform the heteroduplex product into an *ung*+ *E. coli* strain. The nonmutant template DNA strand contains uracil, whereas the mutant DNA strand synthesized in vitro does not. Uracil repair, which occurs as the plasmid begins replication, causes selective loss of the parental nonmutant strand, thereby increasing your chances of isolating the oligonucleotide-directed mutation.

Steps of the procedure are

A. Perform restriction digests of *virD2* (in pCS64) and pUC119

B. Purify DNA fragments from agarose

C. Ligate restriction fragment to vector

D. Transform *E. coli* with the ligated phagemid

E. Prepare plasmid DNA

F. Perform restriction digest of DNAs; examine to confirm insert

G. Prepare single-stranded, uracil-containing DNA template

H. Phosphorylate oligonucleotide

I. Anneal oligonucleotide to template

J. Synthesize DNA in vitro by primer extension

K. Transform synthesis reactions into *E. coli*

L. Prepare plasmid DNA from putative mutants

M. Confirm mutants by restriction analysis of plasmid DNA

Technical Tip

The proportion of mutant versus wild-type transformants will depend on the quality of each step. **Always** design site-directed mutagenesis experiments so that there is an efficient screen for the mutant, such as loss or creation of a restriction site.

Notes

CLONING INTO A PHAGEMID VECTOR

Introduction

You will insert a portion of *virD2* into the phagemid vector pUC119. From the resulting phagemid, you can prepare single-stranded DNA to use as template for oligonucleotide-directed mutagenesis of *virD2*.

Phagemid pUC119 contains the origin of replication for single-strand DNA phage M13; this allows 1 strand of the phagemid (the plus strand) to package into M13 phage particles when a helper M13 phage is infected into the host cells. It also contains a ColE1 plasmid origin of replication, a β-lactamase (*bla;* ampicillin resistance) gene, and a multiple cloning sequence within a portion of the β-galactosidase (*lacZ*) gene. The presence of this portion of *lacZ* will allow you to identify transformants (in your cloning experiment) that harbor a recombinant plasmid with foreign DNA inserted into the multiple cloning site/*lacZ* region. Insertion of foreign DNA into *lacZ* will render the cells Lac−.

LacZ protein normally consists of 1021 amino acids, and the enzyme must form a tetramer to cleave lactose. A deletion mutation, *lacZΔM15,* removes amino acids 11 through 41 and prevents tetramer formation. A peptide (called the α peptide) contains the *N*-terminal 92 amino acids of LacZ, which includes the tetramerization domain. This peptide can interact with the LacZΔM15 protein and thereby restore tetramer formation and enzyme activity. This example of intragenic complementation is usually called alpha complementation. Many plasmid, phagemid, and phage lambda cloning vectors carry the alpha portion of *lacZ* to allow easy detection of inserted foreign DNA. These vectors are used with *E. coli* hosts (such as DH5α) that carry the *lacZΔM15* allele on an F'*lac* plasmid or in the chromosome (usually on a prophage);

note that these strains must not carry a wild-type copy of the *lacZ* gene.

The Lac phenotype of *E. coli* cells can be determined by plating on indicator agar that contains X-gal (5-bromo-4-chloro-3-indolyl-β-D-galactoside) to detect LacZ activity and IPTG (isopropylthiogalactoside) to bind the LacI repressor protein and allow expression of the *lac* operon. When LacZ cleaves the galactoside bond to free the indole moiety, the indole rings react to form indigo, a dark blue compound; before cleavage, X-gal is colorless. Thus, blue (Lac+) colonies contain a phagemid vector lacking an insert, whereas cells that contain a vector with foreign DNA inserted in *lacZ*α will exhibit a Lac− (white) phenotype.

Your starting material is the plasmid pCS64, which contains the entire *virD2* coding sequence flanked by non-*virD2* sequences at each end. Digestion of this plasmid with *Hin*dIII and *Bam*HI will produce one large (vector) fragment, plus fragments of 259, 354, and 560 bp. The target for your oligonucleotide-directed mutagenesis experiment lies within the 354-bp *Bam*HI–*Hin*dIII fragment. To obtain the correct fragment, you will use agarose gel electrophoresis to separate the mixture of restriction fragments, excise the 354-bp fragment from the gel, and use the Geneclean procedure to purify it.

You will ligate the 354-bp *Bam*HI–*Hin*dIII fragment to pUC119 DNA cleaved with both enzymes. Note that the ends of the cut vector are incompatible with each other, so the vector cannot ligate into a circular molecule unless a *Bam*HI–*Hin*dIII fragment inserts into the multiple cloning sequence. Because the insert DNA has 2 different cohesive ends, it will insert into the vector in only 1 orientation. (Had we chosen to insert the target as a *Hin*dIII fragment, it could have ligated to the vector in either orientation.) For oligonucleotide-directed mutagenesis, the orientation of the target sequence relative to the vector is critical, because the mutagenic oligonucleotide must be the complement of the strand that becomes packaged when the phagemid replicates as a single-stranded phage. For the phagemids pUC119 and pUC118, the *lacZ* coding (or sense) strand is also the phage M13 plus (or packaged) strand. We have inserted the *Bam*HI–*Hin*dIII fragment from *virD2* in the opposite orientation relative to *lacZ*: the coding strand of *virD2* is in the minus strand. Therefore, the packaged pUC119–*virD2* single-stranded DNA that you will use as template for the mutagenesis contains the noncoding (or antisense) strand of *virD2,* and the mutagenic oligonucleotide you will use corresponds to the sense strand of *virD2*.

Notes

A. RESTRICTION DIGESTS OF *virD2* (IN pCS64) AND pUC119

Technical Tips

The volume of restriction endonucleases in a restriction digest must not exceed 10% of the total reaction volume because high concentrations of glycerol in enzyme preparations inhibit their activity.

Store restriction endonucleases at −20°C and keep them on ice for the short time you have them out of the freezer. Add them last to a restriction digest mixture, and begin the incubation immediately.

Protocol

1. Mix: 2 μg of insert DNA (the plasmid pCS64)
   2 μL of 10× reaction buffer (NEB *Bam*HI)
   distilled water to give 18 μL total volume
2. Add 10 units each (usually 1 μL) of the restriction endonucleases *Bam*HI and *Hin*dIII. Incubate at 37°C for 1 to 2 hours.
3. Mix: 1 μg of CsCl-purified pUC119 DNA
   2 μL of 10× reaction buffer (NEB *Bam*HI)
   distilled water to give 18 μL total volume
4. Add 10 units each (usually 1 μL) of the restriction endonucleases *Bam*HI and *Hin*dIII. Incubate at 37°C for 1 to 2 hours.
5. Add 1 μL of heat-treated RNase (10 mg/mL).

Solutions for Restriction Digests

1× NEBuffer *Bam*HI: 150 mM NaCl + 10 mM Tris, pH 7.9 + 10 mM Mg Cl_2 + 1 mM dithiothreitol
Heat-treated RNase (10 mg/mL)

Notes

B. PURIFICATION OF DNA FRAGMENTS FROM AGAROSE

Introduction

The *Bam*HI–*Hin*dIII restriction digest of pCS64 resulted in a mixture of fragments; the one containing the target sequence for mutagenesis is 354 bp. It differs sufficiently in size from the other fragments such that it can be separated from the other fragments by agarose gel electrophoresis. Small DNA fragments cannot be separated on the kind of agarose gels you have used so far; instead use NuSieve 3:1 agarose (FMC BioProducts). You will identify the correct fragment by comparison with a size standard; excise it from the gel and use the Geneclean procedure to purify it.

Protocol

1. Prepare a 2% NuSieve 3:1 agarose gel (0.8 g agarose in 40 mL 1× TAE; heat in microwave until dissolved; replace lost volume with ddH_2O; add 1.5 μL of ethidium bromide; pour and insert comb).
2. To the restriction digest of pCS64, add 4 μL 6× load buffer. Load the entire sample into 1 well. Separate DNA by electrophoresis at 100 V (toward positive electrode).
3. Follow the Geneclean procedure, Experiment I.D.

Notes

C. LIGATION OF RESTRICTION FRAGMENT AND VECTOR

Introduction

Inactivation of restriction endonucleases prevents them from cutting restriction sites joined by DNA ligase. For many enzymes, including *Hin*dIII, incubation at 68°C for 15 minutes will destroy the enzyme activity. *Bam*HI is not susceptible to heat inactivation, but can be removed by phenol extraction followed by chloroform extraction and ethanol precipitation. Experience has shown us that *Bam*HI-digested DNAs can be successfully cloned using only a heat treatment before ligation; therefore, the phenol and chloroform extractions are optional. If you choose not to do them, incubate at 68°C instead.

Protocol

Optional:

1. For each restriction digest, bring volume to 0.25 mL with TE (10/0.1). Add 0.2 mL phenol (equilibrated with 1 *M* Tris, pH 8.0); mix well, then add 0.12 mL $CHCl_3$.
2. Mix well, then centrifuge 10 minutes at full speed in microfuge.
3. Save aqueous (top) phase and extract with 0.2 mL $CHCl_3$:isoamyl alcohol (24:1).
4. Centrifuge 1 minute in microfuge; remove aqueous (top) phase to clean 1.5-mL tube.
5. Add 100 μL 7.5 *M* ammonium acetate. Add 0.6 mL of ethanol; mix and hold on ice for 5 minutes.

6. Centrifuge 3 minutes in microfuge.
7. Dissolve pellet in 20 μL TE (10/0.1).

If you choose not to extract with phenol and chloroform, do step 8 instead:

8. Heat DNAs at 68°C for 15 minutes to inactivate restriction endonucleases.
9. Mix: 0.5 μg of linear pUC119
   1 μg of the *Bam*HI–*Hin*dIII fragment from *virD2*
   5 μL of 5× ligase buffer (provided by the enzyme supplier)

   Bring to 25 μL with H_2O.
10. Add 0.5 μL of T4 DNA ligase; incubate at 15°C overnight.

Solutions for Ligation

Phenol equilibrated with IM Tris, pH 8.0

Chloroform

Chloroform: isoamylalcohol (24:1)

7.5 M Ammonium acetate

Ethanol

TE (10/0.1): 10mM Tris, pH 7.5; 0.1 mM EDTA (autoclave)

T4 DNA ligase buffer (NEB):

50 mM Tris (pH 7.8), 10 mM Mg Cl_2,

1 mM ATP, 10 mM dithiothreitol, 25 μg/ml bovine serum albumin

Use ligase buffer provided by the supplier. Store frozen and thaw on ice to protect ATP.

Notes

Notes

D. TRANSFORMATION OF *E. COLI* WITH THE LIGATED PLASMID AND RECOVERY OF CLONES

Follow the protocol for transformation in Experiment I, section F, except use *E. coli* strain DH5α, and plate transformations on LB-ampicillin plates with X-gal and IPTG.

E. SMALL-SCALE PREPARATION OF PLASMID DNA FROM BROTH CULTURES

Follow the protocol for alkaline–SDS plasmid DNA preparation in Experiment I, section G, except add 1 μL of heat-treated RNase (10 mg/mL).

F. RESTRICTION DIGEST OF DNAs: EXAMINATION TO CONFIRM INSERT

Follow the protocol for restriction digestion of plasmid DNA in Experiment III, section A (digestion with *Bam*HI and *Hin*dIII). Examine the restriction fragments on a 2% NuSieve 3:1 agarose gel, with 100-bp DNA ladder as a size standard.

Notes

G. PREPARATION OF SINGLE-STRANDED DNA TEMPLATE

Introduction

For your mutagenesis experiment, you require single-stranded, uracil-containing template. You will prepare this template DNA (pMCB525) in *E. coli* strain CJ236 (*dut-1 ung-1 thi-1 relA1*; pCJ105 [Cm^R = chloramphenicol resistance]).

The M13 phage that produces the single-stranded template infects *E. coli* by attaching to F pili, so you must use a strain of *E. coli* that produces these pili. The F' plasmid pCJ105 carries the *tra* (transfer) genes needed to produce F pili. These pili permit conjugal transfer of plasmid DNA from male (F+ or F') *E. coli* to female (F−) recipient strains. To ensure the presence of pCJ105, we will grow CJ236 in L broth containing 30 μg/mL chloramphenicol.

Unlike most viruses, M13 does not kill host cells. M13 forms turbid plaques by slowing the growth of infected cells. M13 buds from the surface of host cells, so the medium of an infected culture contains virus particles. The M13 capsid contains lipids; therefore, chloroform destroys M13 particles.

Phage titer is measured in plaque-forming units (pfu). Bacterial titer is measured in colony-forming units (cfu). A cfu of 2×10^8 corresponds approximately to a Klett reading of 50 or an absorbance at 600 nm of 0.5 (these values are strain-specific and must be determined empirically). Multiplicity of infection (m.o.i.) is pfu divided by cfu.

Note

Confine RNase to the hood; use designated pipettor.

Phenol is equilibrated with buffer before use, and is usually stored under buffer. Therefore, 2 phases are apparent in the phenol bottle; **do not** mix these. A layer of aqueous buffer is on top. To obtain phenol, insert a pipet into the phenol layer beneath the buffer. After use, collect phenol for organic waste disposal.

Protocol

Beforehand (TAs):

1. Prepare an M13KO7 stock:

   a. Grow a fresh overnight culture of a male (F' or F+) *E. coli* strain (JM101).

   b. Inoculate 2 mL of L broth with 20 μL of JM101 overnight culture and incubate with aeration for 1 hour at 37°C.

   c. Add 20 μL of phage (10^{11} pfu) or one plaque of M13KO7 and incubate 1 hour at 37°C with slow shaking.

   d. Add kanamycin to 50 μg/mL. Incubate with aeration overnight at 37°C.

   e. Centrifuge 5 minutes at 3700 rpm in clinical centrifuge.

   f. Titer phage stock on male strain (JM101). Prepare serial dilutions of phage (in broth). Add 10 μL of each dilution to 0.1 mL of fresh overnight culture of JM101 (or another male strain). Incubate at room temperature for 5 minutes.

   g. Add 3 mL L top agar and plate on L agar plates; incubate 37°C overnight.

2. Grow CJ236 (containing pMCB525) overnight at 37°C with aeration in L broth with 50 μg/mL ampicillin and 30 μg/mL chloramphenicol.

Day 1 (students start here):

3. Inoculate 2 mL of L broth + 0.001% thiamine with 25 µL of overnight culture from step 2. Incubate at 37°C with aeration for 1 hour.
4. Add ampicillin to 25 µg/mL. Incubate at 37°C with aeration for 30 minutes. Titer should reach approximately 2×10^8 cfu at this time.
5. Add 30 to 50 µL of M13KO7 helper phage stock to give an m.o.i. of approximately 20.
6. Incubate at 37°C for 1 to 2 hours with gentle shaking.
7. Add entire 2 mL of infected culture to 8 mL of L broth + 0.001% thiamine + 25 µg/mL ampicillin + 50 µg/mL kanamycin. (The phagemid encodes ampicillin resistance and the helper phage encodes kanamycin resistance.) Incubate with aeration for 16 hours at 37°C.

Day 2:

8. Centrifuge at 10,000 rpm for 10 minutes at 4°C in the Sorval SS34 rotor.
9. Mix 6 mL supernatant with 720 µL 5 *M* NaCl + 960 µL 30% PEG 8000. Hold at 4°C **overnight**; do not skimp on time at 4°C.

Day 3:

10. Centrifuge at 10,000 rpm for 10 minutes at 4°C in the SS34 rotor; discard supernatant and save small beige pellet.
11. Resuspend pellet in 0.5 mL of TE 10/0.1 and transfer to a 1.5-mL microfuge tube. Add heat-treated RNase A to 20 µg/mL. Incubate for 20 minutes at 37°C.

12. Extract twice with 1 volume of a 1:1 mixture of phenol:chloroform (add phenol:chloroform; vortex 1 minute; centrifuge 10 minutes at full speed in microfuge; remove and save aqueous (top) layer, leaving interphase behind). Repeat. Extract with chloroform:isoamyl alcohol (24:1). Thorough phenol extraction is crucial.

13. Add 0.5 volume of 7.5 *M* ammonium acetate, then add 2 volumes of ethanol. Hold on ice for 5 minutes, then centrifuge in microfuge for 5 minutes. Dissolve pellet in 25 μL TE (10/0.1) pH 8.0.

14. Measure optical density (OD) at 260 nm; concentration (μg/mL) = $OD_{260\ nm} \times 40$ (for single-stranded DNA).

15. Examine DNA on an agarose gel to estimate yield of single-stranded DNA. Include a size standard (1-kb ladder) and double-stranded DNA (the plasmid) on the gel for comparison.

Solutions for Single-Stranded Template Preparation

L broth:	Tryptone (Difco)	10 g/L
	Yeast extract (Difco)	5 g/L
	NaCl	10 g/L
	pH 7.0; autoclave 22 minutes	

Ampicillin: 500× stock solution = 50 mg/mL

M13KO7 helper phage stock: titer = 0.5 to 5×10^{11} pfu/mL

Kanamycin: 1000× stock solution = 25 mg/mL

5 *M* NaCl

30% PEG 8000 (polyethylene glycol)

TE 10/0.1 (10 mM Tris, pH 8.0, 0.1 mM EDTA)

RNase A, heat-treated (see Experiment I)

Phenol:chloroform 1:1

Chloroform:isoamyl alcohol 24:1

7.5 *M* Ammonium acetate

Ethanol, 95%

Double-stranded pMCB525 for a size standard

Notes

H. PHOSPHORYLATION OF OLIGONUCLEOTIDE

Introduction

DNA synthesis will extend from the 3' end of the primer, continue around the circular template, and reach the 5' end of the primer. DNA ligase can seal the "nick" between the 3' hydroxyl end of the nascent DNA and the adjacent 5' end of the primer, provided it has a 5' phosphate. Once ligase seals this nick, DNA synthesis cannot proceed farther. If the primer lacks a 5' phosphate, DNA synthesis will continue, thereby removing the mutagenic oligonucleotide. Therefore, you must add a phosphate to the 5' end of the mutagenic oligonucleotide primer before the DNA synthesis step. (Synthetic oligonucleotides lack a 5' phosphate.)

Protocol

1. Mix:

10× polynucleotide kinase buffer	3 μL
1 mM ATP	13 μL
mutagenic oligonucleotide	200 pmol

   distilled water to 30 μL

2. Add 1.5 units of T4 polynucleotide kinase.
3. Incubate at 37°C for 45 minutes.
4. Heat at 65°C for 10 minutes.
5. Add 3 μL of TE (10/0.1, pH 8.0) to dilute oligonucleotide to 6 pmol/μL. Store at −20°C.

Solutions for Phosphorylation

NEB T4 polynucleotide kinase buffer: 70 mM Tris (pH7.6), 10 mM $MgCl_2$, 5 mM dithio threitol

1 mM ATP

200 pmol Mutagenic oligonucleotide: 5' CGC CAG CAG CGA TCT CGC GAT GCT GCG CAA GTT GAT TCC G 3'

T4 Polynucleotide kinase

TE 10/0.1, pH 8.0 (10 mM Tris, pH 8.0, 0.1 mM EDTA)

Notes

Notes

I. ANNEALING MUTANT OLIGONUCLEOTIDE TO TEMPLATE

Introduction

To anneal the mutagenic oligonucleotide to the target sequence, mix them together, heat, and allow to cool **slowly.** As a control, prepare a mixture that lacks the oligonucleotide primer.

Protocol

1. On ice, mix:
   1 μg (0.8 pmol) of your single-stranded template (phagemid) DNA
   1 μL (6 pmol) phosphorylated oligonucleotide primer
   1 μL 10× T4 DNA polymerase buffer
   distilled water to 10 μL.

The primer:template ratio should be 7.5:1.

2. Also on ice, mix:
   1 μg (0.8 pmol) of your single-stranded template (phagemid) DNA
   1 μL 10× T4 DNA polymerase buffer
   distilled water to 10 μL

   (control without oligonucleotide primer)

3. Incubate the tubes at 70°C for 1 minute.
4. Cool to room temperature **slowly** by placing the tubes in a styrofoam float and setting them in a beaker containing 400 mL of 70°C water. Allow the water to gradually reach room temperature.

Solutions for Annealing

Single-stranded template (phagemid) DNA—student preparations

Phosphorylated oligonucleotide primer

NEB T4 DNA polymerase buffer: 0.5 M NaCl, 0.1 M Tris (pH 7.9), 0.1 mM $MgCl_2$, 10 mM dithiothreitol

Notes

Notes

J. IN VITRO DNA SYNTHESIS BY PRIMER EXTENSION

Introduction

DNA synthesis will require single-stranded pMCB525 DNA as a template, DNA polymerase and dNTPs. After DNA polymerase copies the entire template, DNA ligase will join the 3' end of the newly synthesized strand to the 5' end of the primer, preventing further synthesis, which would replace the mutagenic oligonucleotide. Ligation requires ATP.

Protocol

1. To each 10 μL annealed primer/template add:

10× T4 DNA polymerase buffer	1 μL
10 mM ATP	1.5 μL (to give 0.75 mM)
2.5 mM dNTP mix	3.2 μL (0.4 mM final concentration)
T4 DNA ligase	1 μL
T4 DNA polymerase	1.2 μL

2. Incubate 5 minutes at room temperature, then 75 minutes at 37°C.

3. Examine 2 μL of the synthesis product by agarose gel electrophoresis. Include: 1-kb ladder, double-stranded plasmid, single-stranded template, and no-primer control.

Solutions for in Vitro DNA Synthesis

10× T4 DNA polymerase buffer

10 mM ATP

2.5 mM dNTP mix

T4 DNA ligase (400 units/μL; NEBiolabs)

T4 DNA polymerase (3 units/μL)

Notes

Notes

K. TRANSFORM SYNTHESIS REACTION INTO *E. COLI* DH5α

Introduction

Use the following DNAs to transform *E. coli* strain DH5α:

1. Experimental
2. Negative control: no primer synthesis reaction
3. Transformation-positive control: double-stranded pMCB525

Note

Use aseptic technique to avoid contamination of bacterial cultures.

Protocol

1. Transform 2 μL of each synthesis reaction and 1 ng pMCB525 into DH5α according to the protocol in Experiment I.F.
2. Pick 5 putative mutants and 1 colony (pMCB525) from the transformation-positive control; inoculate broth + ampicillin.

L. SMALL-SCALE PREPARATION OF PLASMID DNA

Protocol

Obtain a culture containing a confirmed mutant plasmid. This will be your mutant-positive control. You will prepare

7 DNAs: 2 controls (mutant and transformation) and 5 putative mutants. Follow the procedure in Experiment I.G.

Note: Use RNase. Do 2 ethanol precipitations.

M. CONFIRMATION OF MUTANTS BY RESTRICTION ANALYSIS

Introduction

You will digest plasmid DNA with *Hin*dIII and *Bam*HI and examine samples by gel electrophoresis. The *Bam*HI and *Hin*dIII digestions cleave at the termini of the insert. The parental plasmid will yield one large (>3200 bp) vector fragment and a 354-bp fragment. Mutants will have a 303-bp fragment in place of the 354-bp fragment. To detect these small fragments, use either a polyacrylamide gel or agarose such as NuSieve 3:1 or MetaPhor (FMC BioProducts). We give the protocol for an acrylamide gel; if you prefer, use a 2% NuSieve 3:1 agarose gel.

Protocol

1.	Each digestion:	Master mix:
10 μL	DNA preparation	—
2 μL	10× *Hin*dIII buffer (NEBuffer 2)	16 μL
1 μL	1 μg/μL RNase A solution	8 μL
6 μL	water	48 μL
0.5 μL	*Hin*dIII	4 μL
0.5 μL	*Bam*HI	4 μL

Prepare a master mix (on ice) containing everything except the DNAs. Pipet 10 μL of master mix into 7 tubes, which already contain 10 μL of DNA.

2. Assemble glass plates and spacers.
3. Combine: 1 mL 5× TAE buffer
   2.5 mL 40% acrylamide stock (19:1 acrylamide:bis)
   6.5 mL H_2O
   in a small flask and swirl gently to mix.

   Add: 10 μL TEMED
   40 μL 10% ammonium persulfate
   and swirl gently to mix. Pour gel using automatic pipettor. Insert comb.
4. After gel has polymerized, pipet water onto top of gel, remove comb, and rinse wells thoroughly with distilled water.
5. Place gel in apparatus and add electrophoresis buffer (0.5× TAE).
6. Sample loading:

   Lanes 1–7: 20 μL restriction digest mixed with 5 μL 5× load buffer*

   Lane 8: 5 μL 100-bp DNA ladder, 0.2 μg/μL (premixed with load buffer)

   *Note: use the load buffer containing 2 tracking dyes, bromphenol blue and xylene cyanol.
7. Secure electrodes and apply 150 V (constant voltage) until xylene cyanol (turquoise dye) migrates almost to the bottom of the gel (about 2 hours).
8. Staining: Place gel in tray containing 50 mL H_2O and 5 μL 10 mg/ml ethidium bromide. Stain 5 minutes. Pour stain into collection flask. Rinse gel in distilled water and photograph.

Solutions for Restriction Analysis

NEBuffer 2: 50 mM NaCl, 10mM Tris (pH 7.9), 10 mM Mg Cl_2, 1 mM dithiothreitol

1 μg/μL RNaseA

5 × TAE: 0.2 M Tris-acetate, 0.1 M EDTA

40% acrylamide stock (19:1 acrylamide:bis)

TEMED

10% ammonium persulfate (fresh)

100-bp DNA ladder

5 × load buffer: 0.05% bromphenol blue, 0.05% xylene cyanol, 40% glycerol

10 mg/mL ethidium bromide

EXERCISE 3. STUDY QUESTIONS

1. The sense strand (which corresponds to the sequence of the mRNA) of the region of *virD2* affected by the oligonucleotide-directed mutation is shown separated into codons; bases found in both *virD2* and the mutagenic oligonucleotide are underlined. The sequence below is the mutagenic oligonucleotide. Note that the single-stranded pMCB525 template DNA that you prepared is the complement of the strand shown here.

   A portion of the *virD2* coding sequence:

   5' etc AAG CGC CAG CAG CGA TCA AAA CGA CGT AAT GAC GAG GAG GCA GGT CCG AGC GGA GCA AAC CGT AAA GGA TTG AAG GCT GCG CAA GTT GAT TCC GAG GCA etc 3'

   Mutagenic oligonucleotide:

   5' CGC CAG CAG CGA TCT CGC GAT GCT GCG CAA GTT GAT TCC G 3'

   Is it important for each end of the mutagenic oligonucleotide to anneal stably to the template DNA? Explain.

2. What is the melting temperature (T_m) for each portion of the primer/template duplex?

3. The mutagenesis should create a new *Nru*I site. Where is it?

4. How will this mutation alter the VirD2 protein?

5. Both ends of the primer were designed to form a number of G:C base pairs with the template. Why? Explain why a G:C base pair has greater thermal stability than an A:T base pair.

Notes

EXERCISE

4

DNA Sequencing

Background

Chain termination with dideoxynucleoside triphosphates (ddNTPs) is used in automated sequencers and for sequencing at the bench. Dideoxy sequencing occurs in a set of 4 reactions, each of which contains the DNA template to be sequenced; an oligonucleotide primer complementary to DNA 3' of the area to be sequenced; DNA polymerase; the 4 deoxynucleoside triphosphates (dNTPs), 1 of which is radioactively labelled to allow detection; and a small amount of *1* of the 4 ddNTPs. Either single-stranded template is used or double-stranded DNA is denatured to allow the primer to anneal. During primer extension a ddNTP will sometimes incorporate instead of the corresponding dNTP. Because the ddNTPs lack a 3'-OH, which is required to add the next nucleotide, incorporation of a ddNTP terminates a growing chain at that position. The result is a mixture of DNA chains of different lengths, complementary to the template, each of which was terminated at the position of a specific nucleotide in the sequence. When the 4 reactions are loaded in 4 wells of a denaturing polyacrylamide gel and separated by electrophoresis, the dideoxy-terminated segments form a ladder of bands. The gel is exposed to X-ray film, and the sequence (complementary to the template) is read off the film.

Automated sequencing reactions are labeled with fluorescent dyes, which are detected as they travel past a laser beam during electrophoresis. Four reactions must be run, with 4 different dye labels, 1 for each dNTP. The fluors are attached either to the primer or to the ddNTPs.

You will sequence the putative mutant *virD2* gene created in your site-directed mutagenesis experiment using α-^{35}S-labelled dATP.

Steps of the experiment are

A. Polyacrylamide sequencing gel eletrophoresis

B. Sequencing reactions

C. Automated sequencing

D. Introduction to databases and gene sequence analysis

Safety Precautions

Sulfur 35 can be safely used at the bench. However, once it has been added to the sequencing reaction, all tips and tubes that contact reactions must be discarded in radioactive waste. Line a beaker with plastic wrap and use it to dispose of radioactive trash; empty it into the radioactive waste barrel at the end of the day. When you are done with sequencing, use the Geiger counter to scan your fingers, Pipetman dispensers, lab coat, bench, the floor, and anything you touched, and clean up as needed.

Notes

A. POLYACRYLAMIDE SEQUENCING GEL ELECTROPHORESIS

Introduction

Polyacrylamide sequencing gels contain urea to denature DNA fragments produced during the sequencing reactions. Glass plates for sequencing gels are 20 × 40 cm. The spacers and combs are 0.4 mm thick.

Safety Precautions

Acrylamide is a neurotoxin that should only be handled with gloves. When weighing out dry acrylamide powder to prepare stock solutions, wear gloves, lab coat, and mask.

Unincorporated ^{35}S-dATP will migrate into the buffer in the bottom well. Treat this buffer as radioactive waste.

Technical Tips

Use freshly prepared ammonium persulfate (APS).

Acrylamide begins to polymerize when tetramethyl ethylenediamine (TEMED) and APS are added, so add these immediately before you pour the gel.

Protocol

1. Obtain a set of 2 glass plates. Clean both sides of the plates with ethanol. When dry, assemble plates and spacers. Tape the sides and bottom of the plates with Scotch Brand 3M yellow electrical tape. Put 3 heavy (bulldog) clips on each side of the plates.

2. Mix: 40 mL sequencing acrylamide
32 μL TEMED
140 μL 10% APS

3. Using an electric pipettor, dispense the acrylamide mix into the prepared plates, down one side, at a rate such that the acrylamide does not pool at the top and flows steadily between the plates, which are held almost vertical. Avoid forming bubbles. Fill the gel plates to 1.5 inches below the top. Allow bubbles to rise to the top, then lower to a nearly horizontal position with the top end supported by a rubber stopper. The acrylamide solution will then fill and slightly overflow the glass plates. Insert the flat edge of the comb about 6 mm from the top of the short plate. Place 3 large clips over the comb, pressing the plates firmly over the comb. Allow gel to polymerize about 1 hour. Remove clips over comb, drip a little 1× NNB buffer on comb, or for longer storage cover comb with buffer-saturated filter paper, and wrap with plastic until use. This gel will keep for 1 day.

4. Immediately before use, remove the plastic wrap and filter paper, wash salts and urea off plates, invert comb and reinsert with teeth **just touching** the top surface of the acrylamide. Attach the gel to the electrophoresis device and fill with 1× NNB buffer.

5. Sequencing reactions already contain load buffer. Load 1 μL of each sequencing reaction in the following order: CATGC. Make a record of the loading order. Make sure the gel is not bilaterally symmetrical (so you can tell which side is which when it is exposed to film).

6. Apply 30 W (constant power) until the bromphenol blue dye nears the bottom (slightly less than 2 hours). Turn off power and load a 2nd set of the same reactions (in different lanes). Continue electrophoresis until the bromphenol blue in these lanes nears the bottom.

7. Prepare a tray with enough gel fix in the bottom to cover the gel (approximately 2 cm). Remove gel from electrophoresis device. Slit the tape with a razor and remove it. Remove comb and spacers and **carefully** separate the 2 plates by inserting a metal spatula down one side where the spacer was. The gel will adhere to 1 of the plates. Lower this plate, gel side up, into the gel fix and soak gel for 0.5 hour.

8. Remove gel from the remaining plate by blotting onto 3 MM paper. Dry gel approximately 1.5 hours (dry to touch) in a gel dryer under a vacuum and place in a light-tight cassette on X-ray film.

Solutions for Acrylamide Sequencing Gels

Sequencing acrylamide: 0.5× NNB buffer, 8% acrylamide, 8 *M* urea
For 300 mL, mix 144 g urea, 48 mL 50% acrylamide–2.5% bis-acrylamide, 15 mL 10× NNB buffer, and 126 mL ddH_2O. Filter.
If using purchased acrylamide solution, use 60 mL 40% acrylamide (19:1 acrylamide:bis-acrylamide), 144 g urea, 15 mL 10× NNB buffer; bring to 300 mL with ddH_2O; filter.

10× NNB buffer:
162 g Tris base
27.5 g boric acid
9.3 g EDTA–Na_2
870 mL ddH_2O

Gel fix: 10% methanol, 10% acetic acid, 2% glycerol

Notes

B. DIDEOXY SEQUENCING

Introduction

Sequencing a single-stranded template produces the best results. Single-stranded templates are commonly isolated from clones in M13 or phagemids. However, it is also possible to denature a double-stranded plasmid immediately before sequencing it, for example by treatment with sodium hydroxide or by boiling. The quality of sequence obtained from a double-stranded template depends on the quality of the template; plasmid DNA purified by cesium chloride density gradient centrifugation gives better results than DNA prepared by the small-scale alkaline lysis method. Commercial kits also yield good-quality DNA for sequencing.

You will sequence a double-stranded template, the putative mutant *virD2* gene created in your site-directed mutagenesis experiment. Immediately before sequencing, you will denature the plasmid with sodium hydroxide, neutralize, and precipitate with ethanol. You will use a commercially available kit (Sequenase, U.S. Biochemical) because the balance between the dideoxynucleotides and the deoxynucleotides must be precise.

Sequencing is often done with ^{32}P-labeled primers. These primers are 5'-end-labeled with γ-^{32}P, and they have a short shelf-life. Instead, you will label with α-^{35}S-dATP, which DNA polymerase will incorporate into nascent DNA. (Several kits are available for the non-isotopic detection of sequencing reactions.) You will perform an extra "C" reaction. This makes it much easier to read the sequence.

Safety Precaution

Use gloves and lab coat when handling radioactivity.

Protocol

This protocol is a checklist. Fill out the amounts needed and check off each step as you do it.

1. To 2 to 3 μg of plasmid DNA from a putative mutant, add 1 *M* NaOH to a final concentration of 0.2 *M*. Incubate for 5 minutes at room temperature.
2. Neutralize by adding 0.4 volumes of 5 *M* ammonium acetate (pH 7.5). Mix immediately.
3. Precipitate the DNA with 4 volumes ethanol. Incubate on ice for 15 minutes. Centrifuge at top speed in microfuge for 10 minutes. Wash the pellet in 70% ethanol, air dry, and redissolve in 7 μL ddH_2O. Store on ice.
4. Start 37 and 65°C heating blocks, incubators, or water baths.
5. Prepare annealing mixture:

DNA (approx. 2 to 3 μg)	__ μL
H_2O	__ μL
5× Reaction buffer	2.4 μL
Primer (0.5 pmol/μL)	1.2 μL
Total	12 μL

6. Heat annealing mixture to 65°C for 2 minutes and then cool slowly to less than 35°C for 15 to 30 minutes. Chill on ice.
7. While the annealing mix is cooling, label tubes "C", "C", "A", "T", "G". Fill with 2.5 μL of each termination mix and cap tubes. Place them at 37°C.

ddC/dNTPs	2.5 μL
ddC/dNTPs	2.5 μL
ddA/dNTPs	2.5 μL
ddT/dNTPs	2.5 μL
ddG/dNTPs	2.5 μL

8. Dilute labeling mix 1:10 with ddH_2O. **[TAs will do this step for the class]**

9. Dilute Sequenase enzyme 1:8 in ice-cold enzyme dilution buffer (or TE). **[TAs will do this step for the class]**

10. Labeling reaction: to annealed primer-template, add **in the order listed, making sure to add the enzyme last:**

Dithiothreitol, 0.1 *M*	1.2 μL
Diluted labeling mix	2.4 μL
[^{35}S]dATP	0.6 μL
Diluted Sequenase enzyme	2.4 μL

Mix, incubate at room temperature 2 to 5 minutes.

11. Transfer 3.5 μL of labeling reaction to each termination (ddN/dNTP) tube, mix, and incubate at 37°C for 5 minutes.

12. Stop the reactions by adding 4 μL of stop solution to each tube; store at −20°C.

13. Before loading on a gel, denature at 75°C for 2 minutes.

Solutions for Dideoxy Sequencing

Sequenase kit (U.S. Biochemical, Cleveland, OH)

Sequencing primer: M13 F (universal)

α-^{35}S-dATP

Notes

C. AUTOMATED SEQUENCING

Automated sequencing is frequently carried out using "cycle sequencing", which uses a thermally stable DNA polymerase. Like PCR, the reaction is repeatedly raised to 95°C, allowed to drop to the temperature at which the sequencing primer anneals, and then raised to the correct temperature for the polymerase. This denatures double-stranded DNA, allowing double-stranded plasmids to be directly sequenced. You will use this method to sequence your putative mutant *virD2* and compare the sequence obtained in this way with the sequence you did at the bench.

Notes

D. INTRODUCTION TO DATABASES AND GENE SEQUENCE ANALYSIS

The purpose of this exercise is to introduce some of the options that are available to search genetic databases and analyze nucleic acid and protein sequences. To take full advantage of the laboratory techniques covered in this course, you will need to access genetic databases. You may need to design PCR primers or hybridization probes, or retrieve a gene sequence. You may also need to analyze your own sequences.

It is outside the scope of this course to demonstrate all the computer tools available to molecular biologists, just as in the lab we could not introduce all important protocols. Instead, as in the lab, we will give you a hands-on introduction to a few of them, to give you a place to start and an idea of what can be done.

You have just sequenced a gene. Imagine that you do not know what it is, and you want to identify it. Where do you start?

1. Use Netscape to reach http://www.ncbi.nlm.nih.gov/
2. Select ("click on") BLAST.
3. Select Basic BLAST Search.
4. Leave the Program window set to blastn to search for a nucleotide sequence. Leave the Database window set to nr to search all databases. (If you want an explanation of these options, click the Program or Database buttons. However, **you do not need to do anything here for this project.)**
5. Enter your sequence in the box under Sequence in FASTA format. You can type the sequence by hand, or you can copy a file (from a word processor, for example) and paste it into this box. Enter only nucleotide sequence; numbers and other text are not permitted,

although blank spaces within the sequence do not matter.

6. Click Submit Query.
7. To perform literature searches, use Netscape to reach http://www.ncbi.nlm.nih.gov/ then select Entrez; next, select Literature-PubMed and enter your search terms.

Notes

Notes

EXERCISE 4. STUDY QUESTIONS

1. What is the role of the ddNTPs in sequencing reactions?

2. You perform a sequencing reaction and separate the products by electrophoresis on a polyacrylamide gel, allowing the bromphenol blue to migrate to the bottom of the gel. When you expose the gel to film and develop the film, you see bands only at the bottom of the gel, not at the top. What caused this? What if the bands were at the top of the gel instead?

3. You have just isolated a new thermostable DNA polymerase from a bacterium that grows at 90°C in marine hydrothermal vents. You plan to sequence the gene that encodes this polymerase. What problems might you encounter? Why? Can you suggest approaches to overcome these problems?

4. You loaded the 4 dideoxy reactions in separate lanes on the sequencing gel. In the automated sequencer, the 4 dideoxy reactions were loaded in the same lane of the gel. Why?

EXERCISE

5

Southern Blot Detection of DNA

Background

Southern blots are used to identify a specific restriction fragment (containing a DNA sequence of interest) among a large heterogeneous population of DNA fragments. First, agarose gel electrophoresis is used to separate restriction fragments according to size. Next the DNA is denatured in situ (in the gel) and transferred to a membrane. The DNA is bound to the membrane by UV cross-linking. Finally, the DNA sequence of interest is detected using a hybridization probe, which must be complementary (or nearly complementary) to its target. After a prehybridization treatment to block sites on the membrane where the hybridization probe may bind nonspecifically, the membrane is incubated with the probe. Posthybridization washes remove all probe except that annealed to target DNA bound to the filter. The probe–DNA fragment complex is then detected by autoradiography or a colorimetric method.

To demonstrate the Southern (DNA) blot technique, we will isolate genomic DNA from several different strains of *Agrobacterium tumefaciens* and examine restriction fragment length polymorphisms (RFLPs) in their *virD2* genes. RFLPs are used to estimate relatedness. Using RFLPs from *virD2,* we will group the different *A. tumefaciens* strains by RFLP phenotype.

Steps of the experiment are

A. Prepare genomic DNA

B. Perform restriction digest and agarose gel electrophoresis of products

C. Denature and blot DNA

D. Prepare probe by nick translation

E. Hybridize and wash Southern blot

A. PREPARATION OF GENOMIC DNA FROM *AGROBACTERIUM TUMEFACIENS*

Introduction

Methods for preparing genomic DNA from prokaryotes and eukaryotes (plants, animals, and fungi) share common features. Cells are harvested and then dispersed into a homogeneous suspension; tissues of multicellular organisms must be disrupted, often by grinding. Next, cells are lysed, usually in a detergent–EDTA extraction buffer. Nucleic acids are separated from other cell components, typically by phenol extraction and ethanol precipitation.

Although *virD2* lies on the tumor-inducing (Ti) plasmid in *A. tumefaciens,* isolation of this plasmid is tedious because of its large size (200 kb) and low copy number (1 per chromosome). Therefore, you will isolate genomic (both chromosomal and Ti plasmid) DNA for analysis. You will isolate DNA from the following *A. tumefaciens* strains: A208, A277, A856, ACH5, C58, EHA101, GV3111, K599, A348, A348 with pVK225, R1000, 1855, and 15955.

Technical Tips

Chromosomal DNA is more fragile than the smaller closed circular plasmid DNAs; take care handling and mixing.

Phenol is equilibrated with buffer before use and stored under buffer. Therefore, 2 phases are apparent in the phenol bottle; **do not** mix these! A layer of aqueous buffer is on top. To obtain phenol, insert a pipet into the phenol layer beneath the buffer. Afterward, dispose of phenol properly.

When performing an organic extraction, leave the entire interphase with the phenol phase; you will have to leave behind some of the aqueous phase to avoid debris at the interphase.

Protocol

1. TAs will provide *A. tumefaciens* cultures grown in YEP broth at 28°C with aeration overnight. Each team will receive 4 strains.
2. Centrifuge 1.5 mL of culture for 1 minute at full speed in microfuge.
3. Discard supernatant; resuspend pellet in 480 μL of TE (25/10); add 20 μL of 25% SDS.
4. Incubate 15 minutes at 37°C.
5. Add 57 μL of 5 *M* NaCl; vortex at full power 1 minute.
6. Incubate at 68°C for 10 minutes, then vortex as above.
7. Add 0.5 mL phenol (equilibrated with 1 *M* Tris, pH 8.0); mix well, then add 0.3 mL $CHCl_3$ and mix again. Wear eye protection, gloves, and lab coat.
8. Mix well, then centrifuge 10 minutes at full speed in microfuge.
9. Save aqueous (top) phase and extract with 0.5 mL $CHCl_3$:isoamyl alcohol (24:1).
10. Centrifuge 1 minute in microfuge; remove aqueous (top) phase to a clean 1.5-mL tube.
11. Add 1 mL of 95% ethanol; mix thoroughly and hold on ice for 5 minutes.
12. Centrifuge 3 minutes in microfuge.
13. Discard supernatant; dissolve pellet in 200 μL TE (10/0.1). Add 100 μL 7.5 *M* ammonium acetate.
14. Add 0.6 mL ethanol; mix and hold on ice for 5 minutes.
15. Centrifuge 3 minutes in microfuge.

16. Discard supernatant. Dissolve pellet in 50 μL TE (10/0.1); store frozen.

17. Quantify concentration of nucleic acids (DNA and RNA) by measuring absorption at 260 nm. Dilute samples 1:100 (1 μL DNA in 99 μL TE). $A_{260\ nm} \times$ dilution factor $\times$ 50 = μg/mL.

	Strain	Concentration (μg/μL)	Total yield (μg)
1.			
2.			
3.			
4.			

Solutions for Preparation of Genomic DNA

TE (25/10): 25 mM Tris, pH 8.0; 10 mM EDTA (autoclave)

TE (10/0.1): 10 mM Tris, pH 7.5; 0.1 mM EDTA (autoclave)

7.5 *M* Ammonium acetate: 57.8 g/100 mL (autoclave)

5 *M* NaCl: 29.2 g/100 mL (autoclave)

25% SDS: 25 g/100 mL

Phenol: Melt phenol in distilled H_2O, then add 1 *M* Tris base (12.1 g/100 mL; pH not adjusted) until pH of phenol phase reaches 8.0 (check with pH paper, not a meter); store frozen. Phenol is colorless; pink color indicates oxidized phenol. Quality is important: buy nucleic acids grade, or distill reagent grade yourself.

$CHCl_3$ and $CHCl_3$:isoamyl alcohol (24:1)

YEP broth: 10 g/L peptone (Difco) + 5 g/L NaCl + 10 g/L yeast extract (Difco) (no need to adjust pH). Autoclave 22 minutes

Notes

B. RESTRICTION DIGESTION OF GENOMIC DNA

Technical Tips

Store restriction endonucleases at −20°C. Keep restriction enzymes on ice for the short period that you have them out of the freezer. Add them last to a restriction digest mixture, and begin the incubation immediately.

Protocol

1. Obtain genomic DNAs from different strains from other lab groups. Calculate the volume of each genomic DNA preparation to add to the restriction digestion (add 10 μg total nucleic acids, including RNA).
2. Combine:
   __ μL of genomic DNA preparation
   1.5 μL of 10× reaction buffer (NEBuffer *Eco*RI)
   1.5 μL of *Eco*RI restriction endonuclease (10 units/μL)
   __ μL water to bring to 15 μL
3. Incubate at 37°C for 2 hours to ensure complete digestion.

Solution for Restriction Digest

NEBuffer *Eco*RI: 50 mM NaCl, 100 mM Tris (pH 7.5), 10 mM Mg Cl_2, 0.025% Triton X-100

Notes

C. AGAROSE GEL ELECTROPHORESIS OF RESTRICTION FRAGMENTS

Protocol

1. Mix agarose with 1× TAE buffer (0.8% = 0.24 g/30 mL for minigel); melt in microwave until thoroughly dissolved; add 1.25 μL of 10 mg/mL ethidium bromide.
2. Pour gel into bed with comb (well former) in place; allow to solidify; place in gel apparatus, cover with 1× TAE running buffer, and remove comb.
3. Add 2 μL of 6× load buffer to 10 μL of each sample. Load into individual wells (initially full of 1× TAE). As a molecular weight standard, use *Hin*dIII-cut lambda DNA; **do not use the 1-kb DNA ladder.** Mix 5 μL of a 100-ng/μL solution of the *Hin*dIII-cut lambda DNA ladder with 1 μL 6× load buffer and load at one side of the gel; leave an empty lane between the molecular weight standard and the genomic DNAs.
4. Apply 100 V. DNA will migrate to the positive electrode.
5. When bromphenol blue nears bottom of gel, observe DNA under UV light and photograph. Use the fluorescent ruler in your photograph. Protect gel from contact with the UV light source and other surfaces with plastic wrap.

Solutions for Agarose Gel Electrophoresis

TAE: 0.04 *M* Tris–acetate, 0.002 *M* EDTA
1 L 50×: 242 g Tris base
57.1 mL glacial acetic acid
100 mL 0.5 *M* EDTA, pH 8.0

Ethidium bromide: 10 mg/mL stock

6× load buffer
0.05% bromphenol blue
40% (w/v) glycerol in H_2O

*Hin*dIII-cut lambda DNA (100 ng/μL)

Notes

Notes

D. SOUTHERN BLOT: DENATURATION AND BLOTTING OF DNA

Introduction

During the Southern blot procedure, we denature DNA in the agarose gel, transfer this single-stranded DNA to a nylon membrane, and hybridize it to a labeled DNA probe (*virD2*). The agarose gel contains both large and small DNA fragments. Because large fragments transfer less efficiently than small ones, we use limited chemical hydrolysis to reduce fragment length. Acid hydrolysis partially depurinates DNA (embedded in the agarose gel), and subsequent incubation with NaOH denatures the DNA and breaks the phosphodiester backbone at apurinic sites. Capillary action moves liquid and DNA from the agarose gel upward; the liquid continues through the nylon membrane and into dry paper towels, but the membrane traps DNA. Application of a vacuum or electric current speeds DNA transfer, but requires special apparatus. DNA can be cross-linked to nylon or nitrocellulose membranes by UV irradiation or baking at 80°C in a vacuum for 2 hours. Positively charged nylon membranes do not require cross-linking, but UV irradiation may improve signal and is recommended when blots will be stripped and reprobed.

Protocol

1. Wash the gel for 15 minutes in 0.25 *M* HCl (100 to 200 mL/wash); rinse the gel with distilled water.
2. Wash the gel twice for 15 minutes each in NaOH–NaCl solution; rinse the gel with distilled water after the 2nd wash.

3. Wash the gel twice for 15 minutes each in Tris–NaCl neutralization buffer.

4. Cut Nytran membrane to exact size of gel. Wear gloves and use the liner sheet to keep the nylon membrane clean. Mark one corner of your membrane with a soft pencil.

5. Float the membrane on distilled water in a tray to wet it by capillary action.

6. Soak the membrane in 10× SSC for 15 minutes.

7. Cut 8 sheets of Whatman 3 MM filter paper to the exact size of the gel; saturate the filters with 10× SSC and set 7 sheets on a large piece of plastic wrap.

8. Place the agarose gel on the SSC-saturated Whatman 3 MM paper. Invert the gel so the bottom face will contact the nylon membrane. Use finger pressure (wear gloves) to remove air bubbles trapped between the gel and filters.

9. Lay the nylon membrane on top of the gel, with the pencil mark down. Once the membrane contacts the gel, do not move it, even if the gel and filter are not properly aligned. Use finger pressure to remove air bubbles.

10. Place 1 sheet of SSC-saturated Whatman 3 MM paper on top of the nylon membrane and remove air bubbles. Cover this with a 3-inch stack of dry paper towels (also cut to the same size as the gel). Wrap the entire stack in the plastic film, and set a modest weight on top of the paper towels.

11. Allow DNA transfer to continue for 2 to 16 hours. Transfer is complete when the gel becomes 1 mm thick.

12. Wash nylon filter for 20 minutes at room temperature with 0.2 *M* Tris, pH 7.5, + 2× SSC. Place filter (pen-

cil mark [DNA] side up) on dry Whatman 3 MM paper. Just as the filter begins to dry, irradiate it with 1200 μJ of UV light (using the Stratagene 1800 Stratalinker). This procedure links the DNA permanently to the membrane.

Solutions for Denaturation and Blotting of DNA

0.25 *M* HCl: 12.5 mL of concentrated HCl + 487.5 mL distilled water

NaOH–NaCl solution (0.5 *M* NaOH + 1.5 *M* NaCl): 10 g NaOH + 43.9 g NaCl + 485 mL distilled water

Tris–NaCl neutralization buffer (1 *M* Tris, pH 7.5, + 1.5 *M* NaCl): 157.6 g Tris base + 87.7 g NaCl + 67.7 mL concentrated HCl + 810 mL water

20× SSC (3 *M* NaCl + 0.3 *M* sodium citrate, pH 7.0): 350.4 g NaCl + 176.5 g sodium citrate·2 H_2O + 7.2 mL concentrated HCl + distilled water to final volume of 2 L

0.2 *M* Tris, pH 7.5, + 2× SSC: 24.2 g Tris base + 14 mL concentrated HCl + 880 mL distilled water + 100 mL 20× SSC

Notes

E. PREPARATION OF PROBE BY NICK TRANSLATION

Introduction

Probes may be labeled in vitro by several procedures. The nick translation method uses DNAse I, an endonuclease, to nick double-stranded DNA, and DNA polymerase I to synthesize labeled DNA fragments from these nicks. The random prime technique uses random-sequence 8-base oligonucleotides to prime synthesis of labeled DNA from single-stranded DNA. Many cloning vectors have a phage-specific promoter positioned to transcribe cloned DNA. In vitro transcription from such a promoter, using the corresponding phage RNA polymerase, produces a strand-specific RNA probe.

The most common labels are radioactive (^{32}P) or biotinylated deoxynucleoside triphosphates (dNTPs), or ribonucleoside triphosphates for labeled RNA. Each method of labeling has advantages, disadvantages, and commercial kits to ensure success.

You will use a nick translation kit from GIBCO/BRL to ^{32}P-label pWR160 plasmid DNA, which contains *virD2*. To separate unincorporated label from labeled probe DNA, you will use a Sephadex spin column.

Safety Precautions

Wear film badges when using radioactive isotopes. Wear finger badges **under** your gloves, with the film side turned toward your palm. Work behind plexiglass shields. Use a radiation meter (Geiger counter) to scan your fingers, Pipetman, and work area.

Protocol

1. In a 1.5-mL microfuge tube (on ice) mix:

   1 μg of plasmid DNA (pWR160)
   25 ng of λ DNA
   5 μL of dNTP solution A2 (contains dATP, dGTP, and dTTP; from BRL nick translation kit)
   water to 40 μL total.

2. Add 5 μL of ^{32}P-labeled dCTP.

3. Add 5 μL of DNase I–DNA polymerase I mixture (solution C, supplied with the kit).

4. Incubate at 15°C for 1 hour.

5. Remove the plunger from a 1-mL syringe and plug with siliconized glass wool. Fill with Sephadex G-50 (medium). Or use a commercial spin column, follow the manufacturer's directions, and skip steps 5, 6, and 7.

6. Remove the top of a 1.5-mL microfuge tube and place the tube in the bottom of a 15-mL Falcon tube; place the syringe barrel inside the Falcon tube so that the tip drains into the microfuge tube.

7. Centrifuge 2.5 minutes at 2900 rpm in a Beckman GP (clinical bench-top) centrifuge. Use the GH3.7 swinging bucket rotor. This will pack the column. Fill the void with more Sephadex and repeat the centrifugation.

8. Layer the nick translation reaction on top of the column and centrifuge as in step 7; room temperature is fine. The probe DNA will move rapidly through the Sephadex column and into the microfuge tube; smaller molecules, such as unincorporated label, will be retarded by the pores in the Sephadex beads and remain in the column during the centrifugation.

9. Measure the incorporation with a Geiger counter. To obtain a more accurate assessment, place 2 μL of probe in 4 mL of scintillation fluor and determine the number of counts per minute (cpm) in a scintillation counter. A good probe should have 10 to 100 million cpm/μg of input DNA.

10. Boil the probe for 3 minutes.

11. Chill on ice. Use immediately or store frozen at −20°C.

Solutions for Nick Translation

dNTP mixes in the BRL nick translation kit:
0.2 mM each of the 3 included dNTPs (in 500 mM Tris, pH 7.8, + 50 mM $MgCl_2$ + 100 mM 2-mercapto-ethanol + 100 μg/mL bovine serum albumin)

solution A1: no dATP; contains dCTP, dGTP, dTTP
solution A2: no dCTP; contains dATP, dGTP, dTTP
solution A3: no dGTP; contains dATP, dCTP, dTTP
solution A4: no dTTP; contains dATP, dCTP, dGTP
solution A5: no dCTP, no dGTP; contains dATP, dTTP

DNA polymerase I–DNase I (solution C from kit): 0.4 Units/μL DNA polymerase I, 40 pg/μL DNase I, 50 mM Tris, pH 7.5, 5 mM Mg-acetate, 1 mM 2-mercapto-ethanol, 100 μM phenylmethylsulfunyl fluoride (proteinase inhibitor), 50% (v/v) glycerol,100 μg/ml bovine serum albumin

^{32}P-labeled dATP: [α-^{32}P]-deoxyadenosine 5'-triphosphate; 3000 Ci/mmol, 10 mCi/mL

Nick spin columns: Pharmacia

Sephadex G-50 (medium): Hydrate the Sephadex by steaming for 1 hour in 5 to 10 volumes of 10 mM Tris–1 mM EDTA, pH 8.0.

λ (phage lambda) DNA

Siliconized glass wool: Untreated glass will bind nucleic acids (consider the Geneclean procedure, which depends on this binding). Immerse in 5% (v/v) dichlorodimethylsilane for 5 minutes, rinse thoroughly with distilled water, and dry. Rinse again with distilled water, then bake at 230°C.

Notes

Notes

F. HYBRIDIZATION AND WASHING OF SOUTHERN BLOTS

Introduction

Detection of a specific DNA with a labeled probe involves 4 steps: blocking, hybridization, washes, and detection. Blocking prevents labeled probe, when it is added later, from binding nonspecifically to the membrane. The extent of hybridization between probe and target DNA depends on the "Cot" value; "Co" is the concentration of the free probe at time zero, and "t" is time. To keep Co high, keep the volume as small as practical. Washes remove probe not bound specifically to target sequences. By altering the temperature and salt concentrations of the final washes, we can control the amount of mismatch permitted between probe and target. High-stringency washes (low salt, high temperature) permit fewer mismatches than low-stringency washes.

Technical Tips

Handle blots at an edge with forceps. Bring wash solutions to temperature before use; correct wash temperatures are important.

Protocol

1. Blocking: treat the membrane at 42°C with 10 mL of hybridization solution. Use a sealed plastic tray or a hybridization tube. Agitate gently for 1 hour.
2. Add 100 to 200 ng of denatured probe DNA (10^6 to 10^7 cpm); mix thoroughly. Incubate with gentle agitation

overnight at 42°C. (To denature probe DNA, boil for 3 minutes and then chill in ice water.)

3. Remove hybridization solution and wash the membrane (with constant agitation) twice for 5 minutes with 25 mL of 2× SSC at 42°C.

4. Wash membrane 3 times for 20 minutes (each) with 25 mL of 2× SSC + 1% SDS at 65°C.

5. Wash the membrane 3 times for 20 minutes (each) with 25 mL of 0.1× SSC + 1% SDS at 42°C.

6. Seal the membrane in plastic wrap while it is still damp. This will allow you to later strip the probe from the membrane and rehybridize with a different probe. (If Gene Screen dries completely, the probe may bind irreversibly and delay rehybridization studies.)

7. Under a safelight, load the filter into a film cassette, set a sheet of Kodak XAR X-ray film on top, and place an intensifying screen (Dupont Cronex) over both. Expose the film for 1 to 7 days at −80°C. Exposure time will vary with the specific activity of the probe and amount of probe bound to the target.

   Alternatively, place the plastic-wrapped blot on a Phosphor Imager screen.

Solutions for Hybridization and Washing of Southern Blots

Hybridization solution: 50% formamide, 1% SDS, 1 *M* NaCl, 10% dextran sulfate, 0.01% calf thymus DNA (to make 500 mL: 250 mL formamide + 5 g SDS + 29.2 g NaCl + 50 g dextran sulfate + 5 mL of 10 mg/mL denatured, sonicated calf thymus DNA + 175 mL H_2O to bring to 500 mL)

Calf thymus DNA: dissolve 100 mg DNA in 10 mL distilled water, sonicate for 2 minutes to shear, boil for 3 minutes, then chill on ice.

20× SSC (3 *M* NaCl + 0.3 *M* sodium citrate, pH 7.0): 350.4 g NaCl + 176.5 g sodium citrate·2 H_2O + 7.2 mL concentrated HCl + distilled water to final volume of 2 L

Notes

EXERCISE 5. STUDY QUESTIONS

1. In pictures of the agarose gels showing *A. tumefaciens* genomic DNA digested with *Eco*RI, there are 2 bright bands in most lanes at the lower section of the gel. What do these bands represent?

2. Why does the DNA appear to migrate as a smear rather than as distinct restriction fragments?

3. Before blotting, you soaked this gel in acid and then in base. Why?

4. How could you check whether the DNA transferred to the membrane during the blotting procedure?

5. Which wash is the stringent wash? Why?

6. Will the length of time you perform your stringent wash affect the results? Why or why not?

7. Group the *A. tumefaciens* strains according to their *virD2* RFLPs.

EXERCISE

6

Northern Blot Detection of mRNA

Background

We use Northern blot analysis to estimate the amount and molecular weight of a specific messenger RNA (mRNA) within a preparation of total RNA. The Northern blot procedure measures mRNA accumulation, which reflects both the rate of transcription and the stability of the RNA in question. A hybridization probe can detect a particular mRNA among a population of RNAs separated by electrophoresis and transferred to a membrane.

To demonstrate the Northern blot technique, you will examine expression of the *RbsS* gene in tobacco leaves. This light-regulated nuclear gene encodes the small subunit of the photosynthetic enzyme, RuBisCO. The large subunit of the enzyme is encoded by chloroplast DNA. The RuBisCO holoenzyme catalyzes the addition of CO_2 to ribulose 1,5-bisphosphate, the first step in the Calvin cycle. *RbsS* mRNA is the most abundant message in tobacco leaves, making it easily detectable by Northern blot analysis.

The most difficult part of a Northern blot is preparing undegraded RNA. RNases are ubiquitous and highly stable; they are found in tissue from which RNA is extracted, as well as on human skin. RNases are not destroyed by autoclaving, and are resistant to metal chelating agents. To prevent RNase contamination from fingers, handle all equipment with gloves. During RNA extraction, grind leaves in liquid nitrogen and keep them frozen until they are in phenol or another denaturant.

To eliminate RNases from glassware and solutions, treat with diethylpyrocarbonate (DEPC). Use DEPC at 0.1% (v/v) for at least 10 hours; DEPC and reaction products (CO_2 and H_2O) can then be removed from reagents by autoclaving or heating to 60 to 80°C. DEPC is incompatible with Tris buffer; use MOPS buffer or make Tris buffer from an unopened bottle of Tris and DEPC-treated water. Baking at 230°C eliminates RNases from glassware. DEPC will destroy polycarbonate and polystryrene (e.g., electrophoresis tanks). To decontaminate them, soak in 3% hydrogen peroxide for 10 minutes; remove peroxide by rinsing in DEPC-treated water.

Steps of the experiment are

A. Prepare RNA from tobacco

B. Run agarose–formaldehyde gel electrophoresis

C. Denature and blot RNA

D. Prepare probe by nick translation

E. Hybridize and wash Northern blot

Safety Precautions

DEPC may be carcinogenic; use gloves. DEPC solution will build up pressure in its storage bottle if it contacts water. Always use a clean dry pipet tip when removing DEPC from the stock bottle, and open DEPC stock bottles in a fume hood.

Notes

A. PREPARATION OF RNA FROM TOBACCO LEAVES

Introduction

Methods of isolating RNA must inhibit endogenous RNases and deproteinize the RNA. The most common procedures include either phenol extraction followed by ethanol precipitation, or the use of some other strong denaturant such as guanidine hydrochloride or guanidinium isothiocyanate. When separated by gel electrophoresis, the ribosomal RNA (rRNA) bands provide a visual measure of the yield and purity of the RNA preparation. mRNAs, which have 3' polyA tails, can be separated from ribosomal and transfer RNAs (tRNAs) by oligo(dT) cellulose chromatography.

To prepare total RNA (mRNA + rRNA + tRNA) from tobacco leaves, freeze tobacco leaves in liquid nitrogen and grind with a mortar and pestle. Extract with phenol, and precipitate RNA with lithium chloride.

Freeze and grind 4 leaf samples; each of these will be divided into 3 portions.

Safety Precaution

Use eye and skin protection when using liquid nitrogen and phenol.

Technical Tip

Phenol is equilibrated with buffer before use and stored under buffer. Therefore, 2 phases are apparent in the phenol bottle; **do not** mix these. A layer of aqueous buffer is on top; insert a pipet into the phenol layer beneath the buffer.

Procedure:

1. Chill a mortar and pestle in the freezer.
2. Weigh fresh leaf tissue approximately the size of a quarter and grind to a **fine** powder in liquid nitrogen.
3. Pour into a 15-mL disposable centrifuge tube; solution may spatter, so wear eye protection.
4. Allow liquid nitrogen to boil off completely, then immediately add 2.5 mL RNA extraction buffer.
5. Immediately add 2.5 mL phenol:chloroform 1:1. Vortex for 1 minute.
6. While solution is still homogeneous, divide into three 1.5-mL microfuge tubes; discard remainder.
7. Centrifuge at top speed for 10 minutes.
8. Remove 500 μL of the aqueous (upper) phase from each of the 3 tubes and combine; be careful to leave the interphase behind. Divide into two 1.5-mL microfuge tubes and add an equal volume of 4 *M* LiCl to each.
9. Precipitate RNA overnight at −20°C.
10. Centrifuge tubes at top speed for 10 minutes.
11. Remove supernatant and dissolve pellets in 100 to 200 μL DEPC-treated ddH_2O. Combine.
12. Add 0.4 (sample) volumes of 5 *M* ammonium acetate and 2.5 (final) volumes cold 95% ethanol.
13. Precipitate RNA at −20°C for 2 hours.
14. Centrifuge at top speed for 10 minutes.
15. Decant supernatant. Wash pellet with cold 70% ethanol. Air dry pellet. Dissolve in 20 μL DEPC-treated water.

16. Quantify RNA by measuring absorbance at 260 nm ($A_{260\ nm}$ × dilution factor × 40 = μg/mL). Measure a 1:100 dilution in 100 μL.

Solutions for Preparation of RNA from Tobacco Leaves

Extraction buffer: 100 mM LiCl, 1% SDS, 100 mM Tris, pH 9.0, 10 mM EDTA (do not treat with DEPC; the tissue contains RNase, so the extraction buffer need not be RNase free)

Phenol: (equilibrate with Tris base, pH 7.5; purchase high-quality phenol or redistill it; store frozen at −20°C; should be colorless)

Chloroform

4 *M* LiCl: treat with DEPC, then autoclave

5 *M* Ammonium acetate

70% Ethanol

Distilled water: treat with DEPC, then autoclave

Liquid nitrogen

Notes

B. AGAROSE–FORMALDEHYDE GEL ELECTROPHORESIS

Introduction

Single-stranded nucleic acids such as mRNAs fold into secondary structures (under nondenaturing conditions) that affect their electrophoretic mobilities. Thus we cannot determine the molecular weights of RNA molecules by agarose gel electrophoresis unless the RNA is denatured. Formamide, formaldehyde, and high temperatures are used in sample preparation, electrophoresis buffers, and gels to keep RNAs denatured. Otherwise, agarose gel electrophoresis and blotting techniques are similar to those used for DNA.

Technical Tips

With gloved hands, wipe gel apparatus, combs, gel casting trays, and plastic boxes with RNase Away solution before use. Prepare a supply of baked or DEPC-treated glassware and equipment. To avoid contaminating your RNAs and the RNA size standard, use aerosol-resistant (plugged) tips for your Pipetman.

Procedure

1. Heat a 1% agarose–water suspension (in microwave) until thoroughly dissolved. Hold at 60°C. 1% agarose = 0.5 g/50 mL final volume for minigel. Weigh agarose, using a baked spatula, into DEPC-treated beaker; use a 25-mL disposable pipet to add 43.5 mL of DEPC-treated distilled water; other solutions will be added in the next step.

2. Add 5.0 mL of 10× MOPS buffer and 1.5 mL of 37% formaldehyde, then pour the gel (50 mL total).
3. Combine: 1 to 10 μg of RNA sample (in 5.6 μL)
   2.5 μL of 10× MOPS
   4.4 μL of 37% formaldehyde
   12.5 μL of formamide
4. Incubate at 55°C for 15 minutes.
5. Microfuge sample to bottom of tube, add 5 μL of RNA loading buffer, mix, and load gel. Use 2 size standards: 1) 5 μg of BRL low-range RNA standards, mixed with MOPS, formaldehyde, and formamide and heated as above. **Load in outside lane and leave an empty lane between this marker and RNA**—you will cut this off your gel after it has been photographed; 2) mix 5 μL of a 100 ng/μL solution of the *Hin*dIII-cut lambda DNA with 1 μL 6× load buffer and load in an outside lane.
6. Apply 50 V (5 V/cm) until bromphenol blue dye migrates halfway down the gel; RNA will migrate toward the positive electrode. Electrophoresis buffer is 1× MOPS.
7. Rinse gel 3× with 200 mL water to remove formaldehyde; 6 minutes total.

Solutions for Agarose–Formaldehyde Gel Electrophoresis

10× MOPS buffer: Add 41.8 g MOPS [3-(*N*-morpholino)-propanesulfonic acid] to 800 mL DEPC-treated water; adjust to pH 7.0 (with NaOH or acetic acid as needed). Add 16.6 mL of 3 *M* DEPC-treated sodium acetate, pH 5.2 + 20 mL of 0.5 *M* DEPC-treated EDTA, pH 8.0. Bring to 1 L final volume with DEPC-treated water.

6× RNA loading buffer: 1 mM EDTA, pH 8.0 + 0.25% bromphenol blue + 50% glycerol

37% Formaldehyde

Formamide: To deionize, mix 100 mL formamide with 5 g of AG 501-X8(D) mixed bed resin (Bio-Rad); stir 30 minutes at room temperature, then filter. Freeze at −20°C or prepare fresh daily.

BRL RNA molecular weight markers

*Hin*dIII-cut lambda DNA

RNase Away solution

Notes

C. NORTHERN BLOT: DENATURATION AND BLOTTING OF RNA

Procedure

1. Soak the gel with agitation in 250 mL of 50 mM NaOH for 30 minutes.
2. Rinse the gel with distilled water.
3. Neutralize for 30 minutes in Tris-NaCl neutralization buffer; place the gel in 10× SSC.
4. Cut Gene Screen Plus to exact size of gel. Wear gloves and use the liner sheet to keep the nylon membrane clean. Mark one corner of membrane with a soft pencil.
5. Float the membrane on distilled water in a tray to wet it by capillary action.
6. Soak the membrane in 10× SSC for 15 minutes.
7. Cut 8 sheets of Whatman 3 MM filter paper to the exact size of the gel; saturate the filters with 10× SSC and set 7 sheets on a large piece of plastic wrap.
8. Place the agarose gel on the SSC-saturated Whatman 3 MM paper. Invert the gel so the bottom face contacts the nylon membrane. Use finger pressure (wear gloves) to remove air bubbles trapped between the gel and filters.
9. Lay the nylon membrane on top of the gel, with the pencil mark down. Once the membrane contacts the gel, do not move it, even if the gel and filter are not properly aligned. Use finger pressure to remove air bubbles.
10. Place 1 sheet of SSC-saturated Whatman 3 MM paper on top of the nylon membrane and remove air bub-

bles. Cover this with a 3-inch stack of dry paper towels (also cut to the same size as the gel). Wrap the entire stack in the plastic film, and set a modest weight on top of the paper towels.

11. Allow RNA transfer to continue for 2 to 16 hours. Transfer is complete when the gel becomes 1 mm thick.

12. Rinse membrane in 2× SSC for 5 minutes at room temperature with agitation. Place filter (pencil mark [RNA] side up) on dry Whatman 3 MM paper. Just as the filter begins to dry, irradiate it with 1200 μJ of UV light (use the Stratagene 1800 Stratalinker). This links the RNA permanently to the membrane.

13. Soak membrane in 5% acetic acid for 15 minutes.

14. Soak the membrane in 0.5 *M* sodium acetate + 0.04% methylene blue for 5 to 10 minutes. Rinse with DEPC-treated water until bands appear.

Solutions for Denaturation and Blotting of RNA

50 mM NaOH

Tris-NaCl neutralization buffer (0.5 *M* Tris, pH 7.5 + 1.5 *M* NaCl): 121.1 g Tris base + 87.7 g NaCl + 67.7 mL concentrated HCl + 810 mL water

20× SSC (3 *M* NaCl + 0.3 *M* sodium citrate, pH 7.0): 350.4 g NaCl + 176.5 g sodium citrate·2 H_2O + 7.2 mL concentrated HCl + distilled water to final volume of 2 L

5% Acetic acid

0.5 *M* Sodium acetate + 0.04% methylene blue

Notes

Notes

D. PROBE PREPARATION

Introduction

You will use a nick translation kit from GIBCO/BRL to ^{32}P-label the pUC18-based plasmid, pRbcS, which contains gene for the small subunit of RuBisCO. To separate unincorporated label from labeled probe DNA, use a Sephadex spin column.

Safety Precautions

Wear film badges when using radioactive isotopes. Wear finger badges **under** your gloves, with the film side turned toward your palm. Work behind plexiglass shields. Use a radiation meter (Geiger counter) to scan your fingers, Pipetman, and work area.

Protocol

1. In a 1.5-mL microfuge tube (on ice) mix:
   1 μg of plasmid DNA (pRbcS)
   25 ng of λ DNA
   5 μL of dNTP solution A2 (contains dATP, dGTP, and dTTP; from BRL nick translation kit)
   water to 40 μL total.
2. Add 5 μL of ^{32}P-labeled dCTP.
3. Add 5 μL of DNase I–DNA polymerase I mixture (solution C, supplied with the kit).
4. Incubate at 15°C for 1 hour.
5. Remove the plunger from a 1-mL syringe and plug with siliconized glass wool. Fill with Sephadex G-50

(medium). Or use a commercial spin column, follow the manufacturer's directions, and skip steps 5, 6 and 7.

6. Remove the top of a 1.5-mL microfuge tube and place the tube in the bottom of a 15-mL Falcon tube; place the syringe barrel inside the Falcon tube so that the tip drains into the microfuge tube.

7. Centrifuge 2.5 minutes at 2900 rpm in a Beckman GP (clinical bench-top) centrifuge. Use the GH3.7 swinging bucket rotor. This will pack the column. Fill the void with more Sephadex and repeat the centrifugation.

8. Layer the nick translation reaction on top of the column and centrifuge as in step 7; room temperature is fine. The probe DNA will move rapidly through the Sephadex column and into the microfuge tube; smaller molecules, such as unincorporated label, will be retarded by the pores in the Sephadex beads and remain in the column during the centrifugation.

9. Measure the incorporation with a Geiger counter. To obtain a more accurate assessment, place 2 μL of probe in 4 mL of scintillation fluor and determine the number of counts per minute (cpm) in a scintillation counter. A good probe should have 10 to 100 million cpm/μg of input DNA.

10. Boil the probe for 3 minutes.

11. Chill on ice. Use immediately or store frozen at −20°C.

Solutions for Nick Translation

dNTP mixes in the BRL nick translation kit:
0.2 mM each of the 3 included dNTPs (in 500 mM Tris, pH 7.8 + 50 mM $MgCl_2$ + 100 mM 2-mercaptoethanol + 100 μg/mL bovine serum albumin)

solution A1: no dATP; contains dCTP, dGTP, dTTP
solution A2: no dCTP; contains dATP, dGTP, dTTP
solution A3: no dGTP; contains dATP, dCTP, dTTP
solution A4: no dTTP; contains dATP, dCTP, dGTP
solution A5: no dCTP, no dGTP; contains dATP, dTTP

DNA polymerase I–DNase I (solution C from kit): 0.4 units/μL DNA polymerase I, 40 pg/μL DNase I, 50 mM Tris, pH 7.5, 5 mM Mg-acetate, 1 mM 2-mercaptoethanol, 100 μM phenylmethylsulfonyl fluoride (proteinase inhibitor), 50% (v/v) glycerol,100 μg/mL bovine serum albumin

^{32}P-labeled dATP: [α-^{32}P]-deoxyadenosine 5'-triphosphate; 3000 Ci/mmol, 10 mCi/mL

Nick spin columns: Pharmacia

Sephadex G-50 (medium): Hydrate the Sephadex by steaming for 1 hour in 5 to 10 volumes of 10 mM Tris–1 mM EDTA, pH 8.

Siliconized glass wool: Untreated glass will bind nucleic acids (consider the Gene Clean procedure, which depends on this binding). Immerse in 5% (v/v) dichlorodimethylsilane for 5 minutes, rinse thoroughly with distilled water, and dry. Rinse again with distilled water, then bake at 230°C.

λ (phage lambda) DNA

Notes

E. HYBRIDIZATION AND WASHING OF NORTHERN BLOTS

Introduction

Detection of a specific DNA with a radiolabeled probe involves 4 steps: blocking, hybridization, washes, and autoradiography.

Safety Precautions

Wear gloves and lab coat. Limit exposure to radioactivity. Dispose of radioactive washes properly.

Protocol

1. Prehybridization: treat the membrane at 42°C with 10 mL of hybridization solution. Use a sealed plastic tray or a hybridization tube. Agitate gently for 1 hour.
2. Add 100 to 200 ng of denatured probe DNA (10^6 to 10^7 cpm); incubate with gentle agitation overnight at 42°C. (To denature probe DNA, boil for 3 minutes and then chill in ice water.)
3. Remove hybridization solution and wash the membrane (with constant agitation) twice for 5 minutes with 25 mL of 2× SSC at 42°C.
4. Wash membrane 3 times for 20 minutes (each) with 25 mL of 2× SSC + 1% SDS at 65°C.
5. Wash the membrane 3 times for 20 minutes (each) with 25 mL of 0.1× SSC + 1% SDS at 42°C.
6. Seal the membrane in plastic wrap while it is still damp. This will allow you to later strip the probe from

the membrane and rehybridize with a different probe. (If Gene Screen dries completely, the probe may bind irreversibly and delay rehybridization studies.)

7. Under a safelight, load the filter into a film cassette, set a sheet of Kodak XAR X-ray film on top, and place an intensifying screen (Dupont Cronex) over both. Expose the film for 1 to 7 days at −80°C. Exposure time will vary with the specific activity of the probe and amount of probe bound to the target.

   Alternatively, place the plastic-wrapped blot on a Phosphor Imager screen.

Solutions for Hybridization and Washing of Northern Blots

Hybridization solution: 50% formamide, 1% SDS, 1 *M* NaCl, 10% dextran sulfate, 0.01% calf thymus DNA (to make 500 mL: 250 mL formamide + 5 g SDS + 29.2 g NaCl + 50 g dextran sulfate + 5 mL of 10 mg/mL denatured, sonicated calf thymus (or salmon sperm) DNA + 175 mL H_2O to bring to 500 mL)

Calf thymus DNA: dissolve 100 mg DNA in 10 mL distilled water, sonicate for 2 minutes to shear, boil for 3 minutes, then chill on ice.

20× SSC (3 *M* NaCl + 0.3 *M* sodium citrate, pH 7.0): 350.4 g NaCl + 176.5 g sodium citrate·2 H_2O + 7.2 mL concentrated HCl + distilled water to final volume of 2 L

Notes

Notes

EXERCISE 6. STUDY QUESTIONS

1. Identify bands from the total RNA extract on the photograph of the agarose–formaldehyde gel.

2. Estimate the molecular weight of the RuBisCO mRNA. Measure the distance each molecular weight marker migrated, and plot on semilog paper. Measure the distance RuBisCO mRNA migrated, and use the semilog plot to estimate its molecular weight.

Notes

EXERCISE

7

Protein Interaction Analysis in Yeast

Background

The yeast two-hybrid system is a way to analyze protein-protein interactions in vivo. The system is based on the observation that many eukaryotic transcription factors consist of 2 separable domains: one domain binds a specific DNA sequence, and the other activates transcription. DNA-binding and activation domains from different transcription factors often retain function when combined. The two-hybrid system uses transcription of a reporter gene in yeast to assay the interaction between 2 proteins, one of which is fused to an activator domain, and the other is fused to a DNA-binding domain. If the 2 proteins in question interact, they will bring together the DNA-binding and the transcription–activation domains, and transcription of the reporter gene will occur. If the 2 proteins do not interact, transcription of the reporter gene will not occur, because although the DNA-binding domain can find its DNA target, it requires the activation domain to stimulate transcription.

Two uses of yeast two-hybrid screening are common. First, the system is used to screen for unknown genes encoding proteins that interact with a protein encoded by a known gene. Second, the approach is used to determine whether 2 known proteins interact and to study molecular aspects of the interaction.

You will use the yeast two-hybrid system to determine whether 2 *Agrobacterium* virulence proteins, VirE1 and VirE2, interact. The *virE2* gene is fused to the gene encoding the DNA-binding domain of the repressor protein LexA. The *virE1* gene is fused to the gene encoding the activation domain of the yeast transcription factor Gal4. If VirE1 and VirE2 proteins interact in vivo, the interaction will tether the Gal4 activation domain to a LexA binding site (operator) located upstream from a *lacZ* reporter gene. If *lacZ* is induced, the resulting β-galactosidase will turn the yeast blue in the presence of the chromogenic substrate X-gal (5-bromo-4-chloro-3-indolyl-β-D-galactoside).

You will use a yeast strain that contains the *lacZ* reporter gene. The *virE1* and *virE2* gene fusions reside in separate plasmids. You will transform plasmid DNAs into yeast. As controls, transform with no plasmid, and with the plasmids containing the DNA-binding and activation domains alone, without the fused proteins.

Plasmid	**Contains**
pAD	activation domain of Gal4
pAD–E1	activation domain of Gal4 fused to *virE1*
pBD	DNA-binding domain of LexA
pBD–E2	DNA-binding domain of LexA fused to *virE2*

Steps of the experiment are

A. Yeast transformation

B. Filter β-galactosidase assay

Notes

A. YEAST TRANSFORMATION

Procedure

(TAs do this part)

1. Grow yeast to 10^7 cells/mL in YEPD (1 optical density [OD] at 600 nm = 2×10^7 cells/mL).

(Students start here)

2. Prepare:
   a. Four 1.5-mL tubes, each containing 50 μg denatured salmon sperm DNA (10 μg/μL) and plasmid DNAs.

Tube	Plasmid DNA (0.1 μg/μL)
#1	0.5 μg pBD–E2, 0.5 μg pAD
#2	0.5 μg pBD, 0.5 μg pAD–E1
#3	0.5 μg pBD–E2, 0.5 μg pAD–E1
#4	no plasmid

   b. LITE (100 mM lithium acetate, pH 7.0; 10 mM Tris, pH 7.4; 1 mM EDTA)

   c. PEG–LITE

   d. Selection plates (YNB plus 2% galactose, 30 μg/mL leucine)

3. Harvest 50 mL yeast; centrifuge for 5 minutes, maximum speed in a bench-top centrifuge.

4. Wash yeast twice with TE (10 mM Tris, pH 7.0, 1 mM EDTA).To wash, add a small volume of TE, vortex to resuspend, add TE to 50 mL, and spin.

5. Resuspend yeast in LITE to 10^9/mL (i.e., 0.5 mL if a 50-mL culture was used).

6. Add 100 μL yeast suspension to DNA, mix.

7. Add 0.6 mL PEG–LITE, mix.
8. Shake 30 minutes at 30°C.
9. Heat shock for 15 minutes at 42°C.
10. Spread 100 μL on selection plates (larger volumes reduce transformation efficiency). Incubate 28° C.
    - expect 1000 to 2000 colonies per dish
    - if scoring transformation efficiency, spread 10 μL cells diluted with 90 μL PEG–LITE
    - for a simple plasmid transformation, use 0.5 μg plasmid
    - for a library transformation, use 2.5 μg
    - scale up by using proportionately more DNA, cells, LITE and PEG–LITE

Solutions for Yeast Transformation

50% PEG: Add 25 g polyethylene glycol (molecular weight, 3350) to 50-mL Falcon tube, add sterile water to 50 mL

LITE: 10 mL 1 *M* lithium acetate
10 mL 10× TE, pH 7.0
80 mL sterile water

PEG–LITE: 5 mL 1 *M* lithium acetate
5 mL 10× TE, pH 7.0
40 mL 50% PEG

1 *M* lithium acetate (filter sterilize)

10× TE, pH 7.0: 100 mM Tris, pH 7.0; 10 mM EDTA, pH 8.0 (filter sterilize)

Salmon sperm DNA (10 mg/mL): Purchased, or can be prepared as follows:

Combine 100 mg salmon sperm DNA with 10 mL distilled water. Sonicate 5 minutes at maximum setting.

Extract with 1 volume TE-saturated phenol, then with 1 volume 50:50 (v/v) TE-saturated phenol:$CHCl_3$, then with 1 volume $CHCl_3$. Precipitate with 1 mL 3 *M* sodium acetate and 25 mL ethanol. Wash pellet with 70% ethanol, dissolve in 10 mL sterile distilled water. Store aliquots at −20°C. Before use, boil 5 minutes and chill in ice water.

YNB–galactose agar:
850 mL distilled water
6.7 g yeast nitrogen base (YNB) without amino acids (Difco)
2 g dropout powder
0.1 g NaOH
20 g agar (Difco Bacto agar)

Autoclave, then add:
Galactose to 2% (filter sterilized)
Raffinose to 1% (filter sterilized)

For these experiments, dropout powder should contain adenine (2.5 g), L-arginine (1.2 g), L-aspartic acid (6 g), L-glutamic acid (6 g), L-isoleucine (1.8 g), L-leucine (3.6 g), L-lysine (1.8 g), L-methionine (1.2 g), L-phenylalanine (3 g), L-serine (22.5 g), L-threonine (12 g), L-tyrosine (1.8 g), and L-valine (9 g).

YEPD broth:
900 mL distilled water
10 g yeast extract (Difco)
20 g peptone (Difco)
0.1 g NaOH

Autoclave, then add 100 ml sterile 20% glucose

Notes

B. FILTER β-GALACTOSIDASE ASSAY

Procedure

1. Replica plate onto filter paper:
   a. Fasten a stack of two 12-cm Whatman filters to replica plating block
   b. Place culture upside down on filter and press down firmly
   c. Sharply, lift culture dish from filter (colonies should transfer completely to filter)
2. Trim filter so that it will fit in petri dish lid. Leave tag for label.
3. Freeze filter in liquid nitrogen for 1 minute; thaw for 1 minute.
4. Add 2 mL LacZ–BME buffer and 50 μL 40 mg/mL X-gal to petri lid.
5. Lay 9-cm filter onto puddle of LacZ–BME–X-gal.
6. Lay replica filter, **colony side up**, onto 9-cm filter.
7. Incubate at 30°C.
8. Note color development at 10, 20, 60, and 120 minutes (β-galactosidase-positive colonies will turn blue).

Results

Record the color at the specified times:

Input Plasmids	Color				
	10 min	20 min	30 min	60 min	120 min
pBD–E2 + pAD					
pBD + pAD–E1					
pBD–E2 + pAD–E1					

Solutions for Filter β-Galactosidase Assay

LacZ buffer (store at 20°C)

0.06 *M* $Na_2HPO_4 \cdot 7\ H_2O$	16.1 g/L (8.5 g/L if anhydrous is used)
0.04 *M* $NaH_2PO_4 \cdot H_2O$	5.5 g/L
10 mM KCl	0.75 g/L
2 mM $MgSO_4$ (anhydrous)	0.24 g/L
	Add water to 0.9 L; adjust pH to 7.0; and add water to 1 L

LacZ–BME (make fresh before use)
50 mL LacZ buffer
137 μL β-mercaptoethanol (13 *M* stock)

40 mg/mL X-gal (store at −20°C; TA will supply)
200 mg; dissolve in 5 mL DMSO

EXERCISE 7. STUDY QUESTIONS

1. What is the purpose of each control strain that you tested in this experiment?

2. If you use the yeast two-hybrid protein interaction trap to identify unknown genes encoding proteins that interact with a protein encoded by a known gene, how can you test whether the protein–protein interaction is genuine?

Notes

Index

W

Y